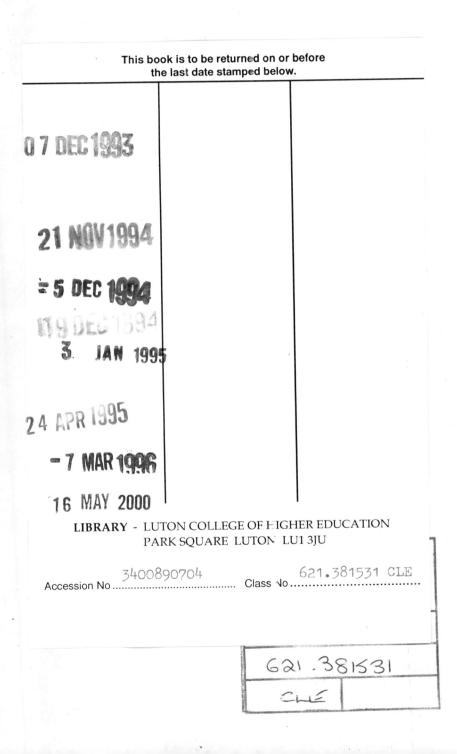

**This book is to be returned on or before
the last date stamped below.**

CLEANING PRINTED WIRING ASSEMBLIES IN TODAY'S ENVIRONMENT

CLEANING PRINTED WIRING ASSEMBLIES IN TODAY'S ENVIRONMENT

Edited by
Les Hymes, P.E.

 VAN NOSTRAND REINHOLD
New York

Copyright © 1991 by Van Nostrand Reinhold

Library of Congress Catalog Card Number 90-43405
ISBN 0-442-00275-0

Manufactured in the United States of America

Published by Van Nostrand Reinhold
115 Fifth Avenue
New York, New York 10003

Chapman and Hall
2-6 Boundary Row
London, SE1 8HN

Thomas Nelson Australia
102 Dodds Street
South Melbourne, 3205, Victoria, Australia

Nelson Canada
1120 Birchmount Road
Scarborough, Ontario M1K 5G4, Canada

16 15 14 13 12 11 10 9 8 7 6 5 4 3 2 1

Library of Congress Cataloging-in-Publication Data

Cleaning printed wiring assemblies in today's environment / edited by Les Hymes.
    p.   cm.
  Includes bibliographical references and index.
  ISBN 0-442-00275-0
  1. Printed circuits--Cleaning.   I. Hymes, Les.
TK7868.P7C56   1990   90-43405
621.381'531--dc20   CIP

FOR MY PARENTS, BROTHERS, AND CHILDREN
AND IN APPRECIATION OF THE SUPPORT
FROM BARBARA

Books are but piles of stone set up to show coming travelers where other minds have been, but no amount of wordmaking could make a single soul to truly know the mountains. One days exposure to the mountains is better than mountains of books.

John Muir
John of the Mountains: The Unpublished Journals of John Muir, p. 94 (University of Wisconsin Press, 1938).

Contents

vii

Acknowledgments

The author acknowledges with gratitude the following companies.

Accel

Advanced Chemical Technologies Inc.

Allied Signal Inc.

Alpha Metals

Baron Blakeslee, Division of Allied Signal

Branson Ultrasonics

Corpane Industries Inc.

Detrex Chemical Industries, Inc.

E.I. DuPont de Nemours & Co.

Electronic Controls Design, Inc.

Electrovert Ltd.

Electrovert Inc.

GAF Chemical Corp.

Martin Marietta Magnesia Specialities Inc.

OSL International Operations Division

Petroferm

Unique Industries, Inc.

Vitronics Corp.

Preface

The impetus to create this book originated from several concerns. One of these was the perceived value to the industry of a collection in one volume of a wide range of information pertinent to the reasons and techniques for defluxing printed wiring assemblies (PWAs). This book is expected to be of use not only to those engaged in the electronics packaging industry but also to those in related fields seeking information concerning viable methods of dealing with one of the environmental issues of our time: the destruction of the ozone layer surrounding and protecting the planet with which we have been entrusted.

The volume of information relative to providing PWAs free of residues adversely impacting operation, reliability, and life of electronic products is growing, and it will continue to expand at an accelerated rate as we seek to match our technology needs and desires with our environmental responsibilities. At the time of this writing, which has spanned the latter portion of 1989 and early 1990, the issue of choosing a new approach to producing PWAs free of detrimental residues while using environmentally acceptable manufacturing techniques appeared to be the major concern of the vast majority of those involved in the printed wiring assembly industry. To many this meant the use of different cleaning media and/or process or equipment enhancements; to others it meant the elimination of the need to clean through materials or process changes. It is to these concerned and responsible individuals engaged in these activities that this book is directed.

The reader will find some approaches, references, and conclusions treated in more than one chapter. One may also perceive some divergence with respect to recommendations offered by the authors. This is to be expected in a compendium of established knowledge and considered expert opinion (as well as bias) covering this field. The differences were intentionally left intact since a totally satisfactory solution for all applications has yet to surface and arrival at a more suitable resolution (or resolutions) of the dilemma may be enhanced by publicizing diverse opinion and objective investigation. The repetitions will be of value in informing those who do not read the total text cover to cover and serve to reinforce the information for those who choose to use the complete book.

We have endeavored to include insights into current options available as well as methods under investigation. It may be that some of the more scholarly would wish a more rigorous treatment and the more practical may desire a definitive single solution, along with more "war stories" upon which to build an approach. We have attempted to present the current state of cleaning technology, knowing full well that time marches on, even as we struggle with the preparation of this book in real time. With this in mind we have also attempted to present some apparently viable paths for the future, in light of the advancing product and process technologies and the concern for the environment.

Les Hymes

1
Design and Process Considerations

Les Hymes

GE Medical Systems, Waukesha, Wisconsin

1.1 OVERVIEW

Production of printed wiring assemblies (PWAs) free of detrimental residues requires consideration of the packaging technology selected, component designs, assembly layout, and the soldering, fluxing, and cleaning processes. Selection of the cleaning agent and process is itself dependent on the chemistry of the total material system. This includes solder, flux, substrate, solder mask, component, and marking materials as well as miscellaneous process materials and contaminants. This chapter will deal with the relationship of some of these design, material, and process issues and their impact on the cleanability of (and the need to clean) the manufactured PWA.

While postsolder cleaning of PWAs is carried out to remove any potentially harmful contaminant from the completed assembly, the principal contaminant of concern normally is the soldering flux residue. Hence, the term *defluxing* is commonly used in place of *cleaning*. PWA design characteristics have a significant impact on the need for flux removal through a direct influence on the quantity of residue remaining and the potential for detrimental reactions. For this reason the chapter will discuss the relationship of PWA packaging characteristics to some defluxing process and material attributes.

Why clean? Postsolder cleaning or defluxing of printed wiring assemblies (PWAs) is often considered a cost adder rather than a value enhancer. However, freedom from harmful contaminants (those adversely affecting desired electronic assembly characteristics and functionality) is as important to reliability over desired product life cycles as the robustness of the soldered interconnections. Ionic contamination above some threshold in humid environments leads to corrosion, electromigration (dendritic growth between conductors in the presence of moisture and electrical bias), and failure of the adhesion of conformal coatings to the substrate and/or component surfaces. In Section 3.1.3 of Chapter 3 of this book Bonner discusses the contribution of ionic species contaminants to these failure modes.

It is important to note that the alternative of using synthetic based low solids fluxes and eliminating the postsolder defluxing process *is* gaining acceptance, particularly outside the USA. The viability of this approach depends on solderability issues, electrical testing impacts, properties of the flux residues, product operating environments and life cycle requirements. The observations made, issues raised, and approaches presented in this chapter relative to PWA design and materials of construction are primarily directed toward manufacturing processes which include a postsolder defluxing step. Design characteristics which increase the possibility of entrapment and the difficulty of removal of aggressive flux residues also increase the concern for the quantity and effect of the low solids/no clean fluxes left in place.

A position can be taken that not all electronic assemblies may need to be cleaned, i.e., the several generally accepted reasons for cleaning presented in Table 1.1 may not apply to all PWAs. The viability of a "no clean" approach depends on component selection, design philosophy, process materials used, handling techniques, and end use for the product.

It has been the prevailing position that PWAs with high I/O count, close spacing, and high reliability, testing, and coating requirements need a postsolder cleaning operation to increase assurance that the total assembly is free of *any* harmful contaminants. This is particularly applicable to those residues left by most of the fluxes in use in 1989/1990.

1.2. ASSEMBLY PACKAGING

The last decade has seen significant evolution in the field of PWA packaging technology. A sizable percentage of the PWAs produced today incorporate surface mount components (SMCs), either exclusively or as a mix with through hole mounted components (THMCs). In addition, the component density exhibited by much of the product using 100% THMCs has also increased. These changes, along with increased circuit densities and finer pitched components, have compounded the difficulty in reaching all areas of the PWA with cleaning agents.

More importantly, the changes have impaired the ability to flush the fluidized mixture of dissolved and particulate contaminants from under and between the components and off the assembly. The more dense circuitry also increases the necessity of removing smaller particles of contaminants in order to eliminate the possibility of electromigration. High speed and high frequency designs, sensitive to the properties of the interconnections, increase the criticality of smaller surface discontinuities on the conductors and the need to avoid even minimal occurrence of dendritic growth. The proliferation of component leads

Table 1.1. The Need for Postsolder Cleaning.

- Remove residues which could contribute to electromigration and result in current leakage between circuitry.
- Eliminate the possibility of corrosion of circuitry and component packages as a result of flux residues.
- Provide for reliable adherence of correctly applied protective coatings by removal of materials which might result in porosity or reduce the bond strength between the coating and the substrate or components.
- Facilitate accurate, reliable, repeatable bed of nails testing and inspection using visual or infrared techniques.

---

and solder joints has been likened to a picket fence hindering the movement of the cleaning agent into and out of the spaces to be cleaned.

It is imperative that the various PWA and component packaging technologies used and the assembly, soldering, and fluxing processes employed be considered when selecting cleaning processes and equipment. This is critical to assuring optimization of the total manufacturing cycle, particularly with changing board level technology.

1.2.1.   Throughhole Mounted Technology

With the increased application of surface mount technology (SMT), the impact of through hole mounted (THM) PWA packaging design on cleaning effectivity has received reduced attention. The assumption is that the issues which exist have been significantly aggravated by use of SMT and any solutions will be similarly applied. Manufacturing and cleaning processes have evolved to accommodate typical THM assembly, layout, and component packaging. Major concerns remaining are similar to those receiving increased attention because of SMT. These concerns includes: (1) conditions which may contribute to potential entrapment of flux residues and cleaning media, (2) compatibility of the materials of construction with the process, (3) the general ability to irrigate the spaces under and around the components.

A few issues which may be more often encountered with THM assemblies include the use of unsealed component packages such as relays, switches, and open winding transformers. These unsealed components are generally unsuitable for immersion in cleaning solutions because of the potential for deterioration of functionality, either through corrosion of contacts as a result of aqueous cleaning or removal of lubricants by solvent or aqueous cleaning processes. These components require mounting after mass soldering and cleaning processes, increasing cost and decreasing the potential for reproducible high quality as a result of hand soldering and cleaning operations.

This situation should be avoided where possible by selection of sealed packages impervious to the cleaning media to be used, or the use of temporary protective covers during the cleaning process. In some cases, a cleaning process limited to the PWA underside is utilized when unsealed components are used. This has the potential for leaving significant quantities of undesirable residues on the topside which have found their way there through component lead holes, vias, other open holes or cutouts incorporated in the PWA design. Similarly, some cleaning media may reach the topside of the assembly in the same manner or through inadvertent overflow around the PWA edges and may remain there, increasing contamination of the topside. Crevices associated with some components (such as connectors) and irregular configurations on the underside of some parts provide sites for entrapment of flux residues. During service under high humidity conditions, these residues may diffuse out onto the PWA surface and contribute to the formation of current leakage paths. These design-related concerns occur in the case of both THM and SM technology and need to be eliminated where possible or addressed in the cleaning process. Items such as stranded wire or wire sleeving can also result in entrapment of flux residues, or can contribute undesirable residues as a result of reaction with the cleaning media. These conditions (as well as flush mounting of dense arrays of dual inline packaged components or pin grid arrays) pose challenges similar to the issues associated with SMT and are discussed in the following section.

During the design stage for either packaging technology conscious consideration must be given to how detrimental flux residues and other potentially damaging contaminants will be prevented from reducing the performance and reliability of the PWA.

Some manufacturers avoid some of these problems through the use of low solids/no clean fluxes which are intended to be left on the completed assembly. These should be considered, particularly where operational or economic factors dictate the use of designs difficult to adequately clean or components which can not be subjected to cleaning media. Before use, evaluation of these less active fluxes should be carried out to assure that the residues left are sufficiently benign so as not to adversely affect the performance of the PWA under the intended service environment.

Potentially damaging residues must be eliminated from the assembly, either through an adequate cleaning process or prevention of their introduction to the manufacturing process.

1.2.2.   Surface Mounted Technology

Compared to THM packaging technology, SMT presents increased difficulty in achieving any specified level of cleaning. Major issues associated with SMT cleaning are design-related attributes which result in significant opportunities for entrapment of residues and impediments to the flow of the cleaning solution. The challenges include: (1) smaller clearances under and around the components, (2) variations in component packaging characteristics, (3) some differences in the materials systems utilized in the process and the product. PWA layout attributes and differences in the component mounting and soldering process employed can also present added difficulties in achieving acceptable cleanliness levels. Among SMT related characteristics are the following:

- Reduced or nonexistent component standoff or clearance from the mounting surface. This interferes with the ability of the cleaning media to penetrate under the device and carry the residues away.
- High component lead count and fine pitch, which reduces the space available between the leads for the cleaning solution to circulate freely to and from the solder joint area.
- Component geometries (including leads on all four sides) which eliminate the ability to orient components to facilitate cleaning solution flow.
- High total component density on the PWA with closer conductor spacing, making removal of undesirable residues more important.
- Configuration under the component, including large numbers of interconnects under large surface area components. The presence of conductors, vias, and adhesives under the components act as obstructions to the flow of cleaning solutions while increasing the probability of the presence of flux residues in this area.
- Substrate wettability and the solubility of residues or contaminants encountered in SMT may vary from the conditions typical of THM assemblies, making effective contaminant removal more difficult.

Process related issues include:

- Reflow soldering process temperature/time profiles can result in an increase in polymerization of some solder paste/flux constituents. This decreases residue solubility in cleaning media and requires more aggressive cleaning processes or materials.
- Flux and adhesive interactions can lead to deterioration of the surface insulation resistance experienced with otherwise acceptable materials.
- Multiple soldering cycles (used when components are mounted on both sides of the PWA) can contribute to cleaning difficulty if the flux residues from the first soldering cycle are not completely removed before the second solder cycle.

At this time the predominant use of SMC is on mixed technology PWAs with these components present in combination with THMC. Use of this type of PWA is growing and it will no doubt be the prevalent second level packaging technology for many applications during the 1990s.

With mixed technology a reflow process is used to interconnect the SMCs mounted in solder paste/flux on the topside of the assembly. The controlled volumes of the solder paste/flux are an aid in reducing the quantity of flux residues. A second soldering cycle is required to interconnect THMCs (leads inserted from the topside) and smaller SMCs (adhesive mounted on the bottom). The second cycle is a wave soldering process requiring the application of a liquid flux to the complete bottom (solder side) of the assembly. As a result of the second mass soldering cycle, the benefit of the control of the flux residue with the reflow process is significantly reduced. Figure 1.1 illustrates a typical mixed technology PWA manufacturing cycle. This mixed PWA packaging technology brings the challenges of both packaging systems to the cleaning operation. In addition to the cleaning difficulties encountered with SMCs attached with the reflow process, the high density of adhesive mounted SMCs on the wave soldered side can trap flux residues in and on any irregularities of the surface of the cured adhesive. The protruding leads and solder fillets of the THMCs can shadow the clearances under the adhesive mounted parts, restricting the access of the cleaning fluid.

In general the use of PWA packaging technology utilizing surface mounted components (exclusively or in combination with throughhole mounted components) increases the difficulty of providing contact between the cleaning media and the surfaces to be cleaned. In combination with the materials system and soldering processes employed, this results in an overall greater challenge to the cleaning process.

1.3. COMPONENT PACKAGING

Component design is the first step in the cleaning process. What the designer chooses with respect to package materials, integrity, and configuration will impact how the PWA to which the component is attached will be mass soldered and cleaned. If process environments are not carefully evaluated these choices may require secondary assembly, soldering, and cleaning operations.

1.3.1. Materials

Compatibility of both the component packaging and marking materials with cleaning media and cleaning process parameters is essential to prevent component damage and/or undesirable contamination of the cleaning media. In order to maintain package integrity, the component packaging materials must be capable of withstanding exposure to the cleaning media without experiencing attack. Reaction with the cleaning media may result in reduction of the ability of the

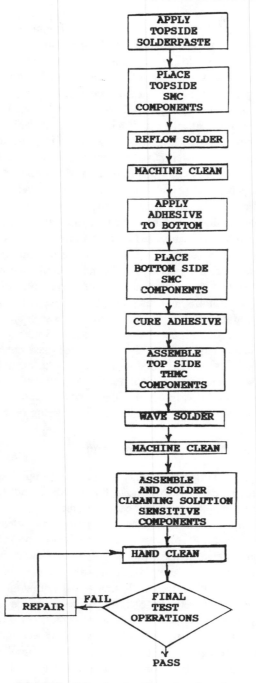

Fig. 1.1. Mixed technology manufacturing process.

component to withstand the intended operating environment. It may also adversely impact the initial functionality by allowing attack of the component internal construction by the cleaning solution.

An example is reaction of the electrolyte in aluminum electrolytic capacitors with chlorine ions present in halogenated solvents which penetrate the package. Epoxy end seals should be employed to avoid this situation. Flux residues on the PWA can attack some surface mount chip capacitors and result in delamination of the structure. Some plastic packaging materials and elastomers have exhibited attack from high power solvents. Polycarbonate and styrene materials in particular can be affected by chlorinated solvents. Exposed contacts in unsealed components are subject to corrosion when using aqueous cleaning if the drying cycle does not remove all liquid from crevices.

In addition to flux residues and other process-related soils, the components themselves may introduce contaminants to the process. These contaminants can include: (1) flux or other residues from lead plating or pretinning operations, (2) package molding die release compounds, (3) mineral spirits or halogenated hydrocarbon fluid residues from component test operations, (4) silicones and light oils from assembly operations, (5) lubricants from unsealed components inadvertently immersed in the cleaning media. Many of these sources of potentially detrimental residues are often ignored when a total PWA manufacturing process is evaluated for cleaning requirements. Introduction of these contaminants to the process requires that the cleaning process to be used be capable of removing them. If left in place there is a potential for PWA degradation in service. If a no clean flux is to be used it is imperative that contaminants other than the flux residue (judged to be benign) be prevented from occurring on the assembly.

1.3.2. Package and Mounting Configuration

Component packaging configurations have a significant impact on the ability of any postsolder cleaning process to remove flux residues and other potentially destructive contaminants. A rough indication of the cleaning difficulty to be expected with a given component is the ratio of package clearance from the substrate to package width. "This aspect ratio compares the access a cleaning medium has to the distance it must travel."[1] The smaller the aspect ratio the more difficult it is for the cleaning media to irrigate the space beneath the component. For comparison a throughhole mounted 16 pin dual inline packaged integrated circuit has an aspect ratio of 0.09, while a surface mounted 16 pin small outline intergrated circuit package has an aspect ratio of 0.05 and a surface mounted 0805 chip capacitor can exhibit an aspect ratio as small as 0.02.

Throughhole mounted disk capacitors are typically mounted to stand above the substrate. Small axial leaded components (generally flush mounted) have virtual line contact with the substrate. These parts present fewer problems in cleaning than large THM dual inline packages, pin grid arrays, and most surface mounted components. The small spacing (between components, between terminations, and between package and substrate), high lead count, and large areas associated with state of the art surface mounted components impede the effectiveness of the cleaning process.

Smaller discrete chip components exhibit very small distances between the terminations and when attached with adhesive, this dimension is reduced even more. Contrary to some opinions the adhesive does not usually prevent flux from penetrating these small spaces unless it completely fills the volume between the terminations, component bottom surface, and the substrate. This condition does not result from common practices. Fluxes are compounded to wet well, and they flow to penetrate the spaces under components and between pads.

Package configuration characteristics can also make it more difficult, or nearly impossible for the cleaning fluid to penetrate the areas typically contaminated. Limiting the volume of cleaning media able to penetrate the space decreases the ability to carry the residues away. The following are several design attributes impacting the effectiveness of the cleaning process and which will be discussed further:

- The height of the component off the substrate surface.
- The total area under the component.
- The total lead or termination count of the package.
- The pitch or spacing between the leads.
- Underside configuration of the package, including the presence or lack of mounting adhesives.
- The presence of crevices or blind pockets, often associated with, but not limited to connectors and sockets.

Table 1.2 presents examples of standoff heights of some throughhole and surface mounted components. Small throughhole mounted axial leaded components (resistors, diodes, and capacitors) are not included. These are generally flush mounted, but as previously noted have virtual line contact with the substrate and generally are not considered an issue. Typical planar areas and lead counts for some of the surface mounted components are also included. All of these characterisitics (standoff height, planar area, and lead count) impact the open volume and shape of the accessible area under the component. These attributes, plus lead pitch, are generally considered the major physical factors of a component that impact the effectiveness of the cleaning operation.

Using a solvent cleaning process, Lermond demonstrated the extent to which smaller standoff heights impact the velocity of cleaning media between glass slides representing component and substrate surfaces. "Increasing the clearance raises the velocity achieved since flow rate through the restricted channel is

Table 1.2. Some Typical Component Standoff Heights.

COMPONENT TYPE	LEAD COUNT	STANDOFF HEIGHT, in.	PLANAR AREA, sq in.
Throughhole Mounted Components			
Plastic DIP	16	0.026	0.19
Ceramic DIP	16	0.021	0.15
Surface Mounted Components*			
Rectangular Chip Components			
RC 0805	2	0.001	0.002
RC 1825	2	0.001	0.045
Leadless Chip Carriers			
(0.050 in. pitch)	16	0.001	0.06
	156	0.001	4.10
SOIC	8	0.007	0.03
	16	0.007	0.06
	28	0.008	0.21
SOT	3	0.004	0.006
Square Plastic J Lead PLCC	44	0.010	0.43
(0.050 in. pitch)			
Quad Flatpack (0.030 in. pitch)	44	0.008	0.18

*Standoff can vary depending on the solder mask used and whether mounted in solder paste for reflow or attached with adhesive for wave soldering.

governed primarily by pressure drop."[2] Musselman has calculated that flow increases as the square of the clearance, and that the resultant drag force on particulate contamination increases as the fourth power of the clearance.[3]

Bonner indicates that with less than 0.005 inch standoff, it was very difficult to remove solder paste/flux residues from under surface mounted leadless chip carriers on PWAs. The results were based on a series of tests using a solvent cleaning system and varying process parameters including spray pressure, conveyor speed, and spray configuration.[4]

With different process parameters and using shimmed glass slides mounted to fiberglass laminate, Willis found that residues were removed from gaps down to 0.004 inch, but that solvent was still present after the samples exited from an inline cleaning system. Testing for residual flux was carried out using a UV lamp technique. The results led to a recommendation that a 0.006 inch clearance would be preferred to facilitate irrigation and reduce the possibility of diluted flux residues being left after evaporation of the cleaning fluid.[5]

It should be mentioned here that an increase in standoff height, although an aid to cleaning, is not always an available alternative, since the resultant increase in inductance in the circuitry may not be acceptable, particularly, for high speed, high frequency applications. Although substrate pads provide some measure of standoff, they also are a distinct and significant barrier to effective flow of the

Table 1.3. Impact of Lead Pitch on Spacing Between Substrate Pads.

COMPONENT TYPE	PITCH P, in.	TYPICAL DISTANCE X BETWEEN SUBSTRATE PAD EDGES, in.
Leaded THM DIP	0.100	0.045
Surface mount components	0.050	0.025
	0.040	0.020
	0.030	0.015
	0.026	0.012
	0.020	0.008

cleaning media. With a constant component perimeter, as the lead count increases, the pitch or distance between the interconnects is reduced. This in turn reduces the opening available for communication of the cleaning fluid with the area under the component. The reduced opening emphasizes the characteristics of tubular capillarity which governs the flow of the fluid under most components. Table 1.3 provides an indication of the typical space between the edges of substrate pads with various component termination pitches in use today.

With finer pitch, not only is the difficulty of cleaning increased, but the criticality is also increased since reliable functionality of more closely spaced circuitry is less tolerant of any type of residue.

All of the component configuration attributes mentioned in this section impact the ability of the cleaning solution to irrigate the space under the component and remove residues. The volume of a *suitable* cleaning solution made available to the sites of the residues is a governing factor in achieving adequate cleaning. This volume must be sufficient to solubilize the residues and flush away reaction products and particulate contaminants. The characteristics of the residues encountered dictate the volume required. The more tenacious and insoluble the residue the more volume required. Characterization of the flux and other residue sources with respect to the processing planned is critical in determining the compatibility of the materials and processes.

Wang and Seghal have described the spacing beneath surface mount components as capillary configuration.[6] As illustrated in Figure 1.2, components

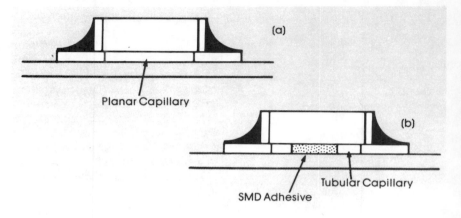

Fig. 1.2. Capillary geometries of surface mounted components: (a) mounted in solderpaste and reflow soldered, (b) mounted with adhesive and wave soldered. (Reprinted with permission from *Printed Circuit Assembly,* August 1988, p. 16, Fig. 6.)

mounted in solder paste and reflow soldered will exhibit a planar capillary geometry; while those mounted with adhesive before wave soldering have tubular capillary clearances. These geometries require consideration of capillary flow phenomona in addition to contact angle and wetting characteristics. Capillary action (the ability of a fluid to penetrate small gaps) is driven by the pressure differential across an interface or meniscus. In addition to Wang and Seghal, others indicate that the driving force can be calculated to be *directly* proportional to the surface tension.[7] With this in mind, it should be apparent that there is a solid technical basis for concern with the component characteristics discussed in this section. The area under the part, the spacing to the substrate, the lead count, and the pitch all have an impact on capillarity.

No consideration is given here to distinctive properties of specific residues or the rates of reaction between any specific cleaning solution and the residues. The intent of this section is to illustrate the impact of some component packaging characteristics on the fluid dynamics of the cleaning process without regard to the type of cleaning solution used. It is left for authors in other parts of this book to discuss and interpret the properties and suitability of the cleaning materials and processes involved.

With the varied and significant impacts of component packaging on the effectiveness of any post soldering cleaning process, it is critical that design engineers as well as process engineers are acquainted with the characteristics of components selected. The continued upward trend in density at all levels of electronic packaging, along with increased constraints on the use of solvents, will add to the requirements for innovation in component packaging as long as the need for defluxing continues to be required. Because of component features and PWA characteristics, as well as environmental constraints, higher power cleaning chemistries may not be the answer. Sources of mechanical energy assist to the cleaning process may prove more suitable. One area of concentration will be component constructions that are capable of withstanding ultrasonic augmented cleaning processes. Increasing interest is apparent in the use of ultrasonic energy as an assist in irrigation of tight spaces to displace flux residues. Continuing studies at the Electronic Manufacturing Productivity Facility operated by the U.S. Navy at Ridgecrest, California indicate that some current component packages are compatible with exposure to ultrasonic energy of specified frequency and power levels for reasonable periods of time.[8]

The previously mentioned alternative of using no clean fluxes requires that these materials be adequately tested to assure demonstration of the lack of adverse effects in the proposed service environment. Properties of some of these fluxes and resultant residues are treated in detail in Chapter 2 by Dr. Guth.

1.4. THE PRINTED WIRING ASSEMBLY

The PWA layout has significant impact on the task of postsolder cleaning. There is a potential for loss in product functionality and long-term reliability if the layout does not exhibit characteristics compatible with the capability of the proposed cleaning process to remove potentially destructive residues. Detailed consideration must be given to the PWA layout at the outset of the design process if effective and efficient operation of a postsolder cleaning system is to be achieved.

1.4.1. Substrate and Solder Mask

The structure and materials of construction of the substrate and solder mask will have an impact on the suitability and the effectiveness of the cleaning process selected.

Substrate materials generally can withstand the cleaning agents currently in use provided that proper curing of the substrate takes place during fabrication and the cleaning process incorporates a suitable drying step. Particularly when a soldermask is *not* used, the designer and process engineer should be fully aware of the substrate material properties with respect to the impact on the selection and control of the cleaning process. The materials of construction, substrate fabrication processes, and soldermask impact the base surface insulation resistance (SIR) exhibited by a PWA. Differences in solder mask and the extent of curing achieved during processing can affect the SIR and impact PWA long-term reliability in some environments.

The representative SIR level of the unpopulated substrate construction should be known before attempting to assess the effectiveness of any PWA postsoldering defluxing operation.

Design dictated cutouts in the PWA, which allow wavesoldering fluxes un-restricted access to the topside of the board, increase the possibility of leaving flux residues entrapped under components. These features should be avoided where possible. Length, width, thickness ratios requiring the use of stiffeners during wavesoldering also provide the potential for entrapping flux between the stiffener and the assembly in areas where removal may be difficult.

The impact of the properties and integrity of a solder mask are significant to the cleaning process since when used the solder mask may account for 60% or more of the surface area of the PWA. Tight, complete, and dependable adhesion of a nonporous solder mask is critical in preventing entrapment of soldering flux residues and other contaminants between the mask and the substrate where contact with traces or terminations could cause circuitry damage. Solder mask applied over solder coated traces and/or large groundplane areas can "wrinkle" and crack during mass soldering. This is a result of reflow of the solder beneath the mask. This can lead to crevices for residue entrapment. Where this is a potential problem solder mask over bare copper should be considered.

Materials left from wet processes involved in substrate fabrication can also be a source of residues on the PWA. When no soldermask is used, these residues are difficult to remove in the postsoldering cleaning operation. In cases where solder mask is used and these residues do not come in contact with the cleaning media, they remain in place as a potential source of deterioration of the long term reliability of the assembly through corrosion, electromigration, or deteri-oration of the functionality of the solder mask.

Photoimageable solder masks, both dry film and liquid, have achieved prom-inence as preferred products for state of the art, high density PWAs with fine line circuitry. With any solder mask the flux chemistry and solder mask inter-actions must be considered when selecting a cleaning process. Activators and other constituents of the proprietary flux compositions may react with partially cured areas of the solder mask.

The presence of incompletely cured solder mask is one cause of the occurrence of solder balls as residues. Solder balls which may occur for other reasons, have a tendency to stick tightly to incompletely cured areas of the solder mask. Agitation and cleaning solution velocity are of assistance in removing these particulate residues.

As indicated in Figure 1.3, solder mask effectively reduces the clearance under the mounted components, increasing the difficulty of flushing out the residues. Solder mask thicknesses can range from 0.006 inch to less than 0.001 inch. There have been suggestions made to increase the standoff height by using thick adhesive deposits for bottomside mounted components. This approach then de-pends on the solderwave to form a larger fillet. Use of this technique should be

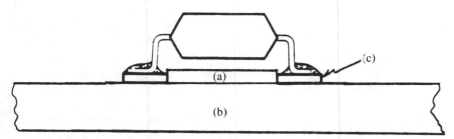

Fig. 1.3.   Representation of the effect of soldermask on clearance under surface mounted components: (a) soldermask, (b) substrate, (c) substrate pad.

investigated in each specific application for initial quality and long-term solder joint reliability tradeoffs. Thought should be given to the possibility of negative impacts on the ability to form a satisfactory solder joint in the presence of a large gap between the substrate land and the component termination.

1.4.2.   Component Placement

Placement of components on the substrate needs to be examined carefully during the design and layout stage, as much for cleanability as for soldering performance and interconnect routing. There are tradeoffs and compromises in most cases. During design producibility reviews it is essential that the impacts of component orientation, spacing, trace routing, and via location on cleanability are understood and resolved.

Component placement strategies which are beneficial to good soldering performance are, in general, satisfactory for cleanability efficiency. However, the flow of the cleaning solution across the surface of the assembly may not always be parallel to either the movement of the PWA through inline cleaning equipment or the largest opening under the component. In the case of batch cleaning systems, the racking of the assemblies has a major impact on the direction of the flow of the liquid media. Physical agitation achieved by mechanical and spray techniques aids in overcoming some of the orientation impacts. Higher component densities and smaller clearances from the substrate increase the importance of incorporating consideration of cleaning issues early in the design stage, well before the component placement activity in the CAD area.

Gruenwald and Lowell[9] evaluated the cleanability of various surface mounted components on three solder mask types in a controlled study of aqueous and solvent cleaning processes. They selected components to represent a range of areas and capillary openings. Placement was varied to offer orientations both perpendicular and parallel to cleaning solution flow. The authors concluded that various solder masks affected cleaning efficiency differently as a result of thickness, wetting characteristics, and propensity to retain flux residues in the surface of the mask. Different base surface insulation resistance values were obtained with unpopulated boards, indicating that knowledge of this effect is indeed necessary for good evaluation of results. Cleaning efficiency using high velocity sprays appeared insensitive to component orientation. An exception was the case of a large rectangular chip resistor mounted so that the solution flow impinged on the end of the component rather than the side. In this case less than satisfactory residue removal resulted. However, during later communication with the investigators, it was stated that subsequent modification of the cleaning equipment resulted in satisfactory cleaning results with the larger component in the original orientation. These studies were conducted using an IR reflow soldering process and a non-halide bearing flux. Direct extrapolation of the data to other processes is not advised since cleaning equipment characterisitics can impact the results.

Component leads and other projections entrap flux and interfere with the flow of cleaning solutions, reducing cleaning efficiency. This applies to the terminations on both leadless and leaded surface mount components and throughhole mounted component leads. During the PWA design layout stage, component placement relative to the directionality of the cleaning media flow across the PWA surface should be considered. Staggering and spacing the components in order to avoid obstructing solution impact and flow should be attempted wherever possible.

Although sometimes used to conserve board real estate, the mounting of chip components under plastic leaded chip carriers is not recommended from a cleanability standpoint because of the obvious reduction in clearance and the introduction of solderpaste flux in spaces difficult to irrigate. Likewise, the placement of traces and vias (plated through holes without component leads, used as interlayer connections) under components makes cleaning more difficult. The traces form obstructions to the flow of the cleaning media under the component and can effectively dam the area, retaining flux and preventing contact with the cleaner. The solder mask over the traces and the thickness of the trace itself can decrease the clearance by as much as 0.006 inch.

The vias allow flux from wave soldering operations to flow up under the component and enter sites where contact with cleaning media is restricted. Where the via cannot be placed in a configuration fanned out from under the component, tenting of the via with solder mask is one accepted method of eliminating this concern. When tenting is used, it must be done with good understanding of the solder mask thickness and properties. Thick masks reduce the clearance, while masks with inadequate strength can rupture and become locations for entrapment of residues.

Wherever they are located, decreases in trace width and spacing increase the criticality of assuring that any residues which may ultimately result in long term corrosion or electromigration are removed during cleaning. Removal of very small residue deposits or solder balls becomes increasingly important with fine-line/fine pitch technology. ANSI/IPC-S-815B "General Requirements for Soldering Electronic Connections" requires for high reliability and general industrial applications that solder balls remaining after cleaning be a maximum of 0.005 inch diameter and not violate the minimum trace spacing. Some investigators advise that residual particles greater than 20% of the intertrace spacing must be removed to assure high reliability.[10]

1.5 ASSEMBLY AND SOLDER PROCESS

The impacts of electronic packaging design and process specification on post solder cleaning are manifested throughout the PWA assembly and soldering operations. In order to assure reliable end product operation, cleaning technology considerations must be an integral part of the design and manufacturing process selection. An understanding of the origin and nature of contaminants introduced during the total manufacturing process required for a given design is essential to selection and control of an effective cleaning process.

1.5.1. Assembly Aids

One possible source of contaminants introduced during the manufacturing process are materials used to aid the process, but which do not remain as part of the end product. These include temporary solder masks used to mask specific areas from solder deposition during wave soldering. Temporary solder masks include tapes and compounds intended to be soluble in the cleaning media. For solvent cleaning processes these materials are often wax-based. For aqueous cleaning they are compounded of gums, thickeners, and alcohol. Taping materials used to hold components on reels for machine insertion or placement can contribute adhesive residues to the component package. Knowledge of the materials to be encountered is necessary to cleaning process control.

1.5.2. Component Mounting Adhesives

Adhesives are used to attach SMT components to the surface of the printed wiring board that will contact the solder in a wave soldering process. The adhesive materials used need to be controlled to assure the capability of withstanding the soldering and cleaning operations without deterioration and without formation of residues which are difficult to remove and potentially detrimental to the end product operation. IPC-SM-817 "General Requirements for Dielectric Surface Mounting Adhesives" provides information and test methods for aid in selection and control of these materials.

Use of adhesives requires control of the size and configuration of the deposit in order to avoid impeding the flow of the cleaning media under the component and entrapment of residue-laden fluids. The deposits have been considered an aid in enhancing irrigability by increasing the standoff of the component from the assembly surface; yet, in many cases, they complicate the cleaning process by acting as a barrier to contact of the cleaning fluid with the flux residue contaminated areas.

Other adhesives and electrically insulating or thermally conductive materials are sometimes used between components and the substrate or heatsinks. The composition of these compounds should be evaluated for cleaning system compatibility and potential for replacement with other suitable methods of accomplishing the desired end.

In general, evaluation should be carried out to assure that materials added to the PWA, and residues of those intended to be removed, do not deteriorate the desired surface insulation resistance level of the completed assembly.

1.5.3.   Soldering Cycles and Fluxes

When considering the cleanability of the product, the fluxing, preheat, and soldering cycles of the PWA manufacturing process present significant challenges for the design engineer and the process personnel. Impacts of the choice of flux type and method of application include:

• Increased quantities of residues from use of high solids content and/or highly activated fluxes. These materials are often selected for two reasons: (1) to enhance soldering process performance when using components with doubtful, uncontrolled, or poor solderability properties, and (2) to increase throughput rates with dense or high heat capacity assemblies. Residues left from use of these fluxes tend to pose increased potential for problems with degradation of PWA functionality.

• Application of excessive amounts of flux because of less than adequate control or intrinsic properties of the application process selected.

• The potential for reaction of the flux with the substrate or solder mask (resulting in a porous surface with absorbed or reacted flux residues difficult to remove). This can occur when the substrate/solder mask chemistry is not considered when selecting a flux. Inadequate curing of the soldermask also contributes to this condition.

• Use of more benign, less aggressive fluxes requires components and printed wiring boards with excellent solderability. It requires the controls and dicipline necessary to obtain parts with the necessary level of solderability and maintenance of that characteristic up to the point of soldering.

• When applied in controlled and limited quantities, low solids/no clean fluxes appear to exhibit the benefits and problems of a benign flux. Components which are sensitive to cleaning operations can be included in the mass soldering op-

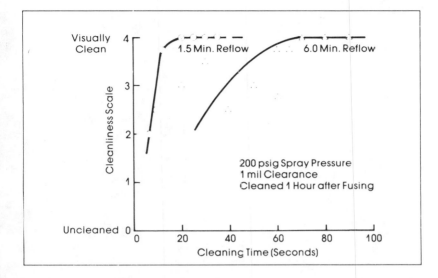

Fig. 1.4. The effect of reflow time on time to visually clean aged residues. 0.001 inch clearance, 1.5 minute reflow time. (Reprinted with permission from *Printed Circuit Assembly*, May 1987, p. 24, Fig. 14.)

eration when using these no clean fluxes since no postsolder cleaning operation is required. This eliminates the necessity of a second, handsolder operation. A problem encountered with many current fluxes of this type is the possibility of interference of the residue with bed-of-nails electrical test and infrared solder joint inspection techniques. Newer low solids fluxes are under development in an attempt to improve testability (see Guth, Chapter 2 of this book).

The type of soldering process and the specific conditions employed have distinctive effects on the properties of the residues which must be removed. These effects include:

• Solder paste fluxes used in reflow soldering processes tend to produce residues which are more difficult to remove than the residues of liquid fluxes associated with the wave solder process. This is attributed to the presence of thixotropic agents and plasticizers, flux activators, and the parameters of the reflow process. High rates of heating and longer times at elevated temperatures associated with the reflow processes can cause increased isomerization of the flux and a more stable, viscous, tenacious, less soluble residue.[11] Figure 1.4 illustrates the impact of increased reflow time. This effect is also noticed with liquid fluxes in the wave soldering process when excessive preheating rates and temperatures were used.

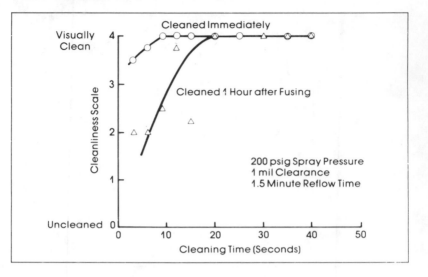

Fig. 1.5. The effect of aging on time to visually clean residues after 1.5 minute reflow time. (Reprinted with permission from *Printed Circuit Assembly,* May 1987, p. 24. Fig. 13.)

• With infrared reflow solder operations it is possible for flux originating in the solder paste deposit to fill completely the space under small components or stay near the warmer edges of flush-mounted components, resulting in a "dam" which must be breached in order for the cleaning media to penetrate.[12] When this happens with flux applied during wave soldering, much of the excess flux is displaced by the dynamics and high volume of solder present with this process. However, based on modeling it has been reported that even with the wave solder process, the residue can completely fill the space beneath the component. To remove the residue it then requires a high pressure spray (or other form of added energy) to form a channel to allow irrigation of the space.[13]

• With vapor phase reflow soldering, it is possible for solder paste fluxes with high rates of solubility in specific fluids to deposit out on portions of the equipment and surfaces of the assembly. This results in the potential for more widespread contamination with a tenacious residue.

• When multiple soldering passes are used (as is common with mixed technology assemblies) it is advisable, if not mandatory, to clean between the soldering operations. Failure to do so increases the carbonization/isomerization of the residues from the first fluxing/soldering operation. Similarly, delay in performing a cleaning operation after soldering can increase the time to obtain residue removal. Figure 1.5 illustrates this condition. Comparison with the previous

Figure 1.4 indicates that extended reflow time has a greater impact than an increase in time between soldering and cleaning. However, there appears to be an interaction between the two effects and the time from reflow to cleaning should be kept to a minimum. With reflow times typical of current manufacturing processes, increased aging time increases the difficulty of removing residues. Lermond indicates little difference in cleanability of samples subjected to either 1.5 or 6 minutes reflow cycles with a 2 minute delay before cleaning. A 15 minute delay also had little effect on cleanability after the 1.5 minute reflow but significantly increased the time to visually clean those samples exposed to a 6 minute reflow.[14]

• Hand soldering, after mass soldering and cleaning, may be required as a result of component selection, other characteristics of the PWA, or rework/modification operations. When this is the case, it is essential that care be taken to limit additional contamination of the cleaned assembly. It is recommended that the hand soldering be performed with as benign a flux as possible. Avoid highly activated fluxes. The real need for this added operation should be carefully evaluated. Design requirements, ineffective soldering processes, or specification of rework for cosmetic reasons should be addressed by other means whenever possible. Spot cleaning is, at best, a marginally effective operation, sometimes adding contamination or spreading residues into critical areas.

1.5.4. Conformal Coating

An effective cleaning process and good manufacturing procedures for material handling and control of airborne contaminants are critical to the adhesion and subsequent performance of a conformal coating. These coatings are applied to protect the assembly from environments predicted to be aggressively destructive to continuous and reliable operation of the PWA. However, the coatings serve to seal contaminants in as well as out. Coating over ionic residues may result in corrosion of the PWA circuitry. Coating over residues can also degrade the adhesion and result in an ineffective encapsulation. Fingerprints left during inspection, test, and miscellaneous handling can be a source of ionic contamination *after* a cleaning operation and before the final coating operation. Attention to detail, handling methods, and workplace environment are essential to maintaining the acceptable level of cleanliness obtained with the final cleaning operation.

1.6. CLEANING OPERATION

PWA cleaning processes currently in use, as well as those under development, focus on the ability to move fluids across the surface of an assembly and through the spaces where residue is located. The mission is to transport solubilized and particulate contaminants off of and away from the PWA. Properties of the cleaning solution and energy added by the process in the form of heat and mechanical agitation are used to enhance and facilitate an effective cleaning process.

An efficient and effective production line cleaning operation requires close and continuous attention to control of process parameters. The difficulties encountered in cleaning state of the art PWAs are also present with techniques used to assess the effectivity of the process.[15] Some measurement techniques used to determine the cleanliness level of the PWA depend on penetration and irrigation of the same small spaces which present a challenge to the cleaning process. For more detailed treatment of these measurement techniques in this book, refer to Munie, Chapter 4, Section 4.4.2 and Bonner, Chapter 3, Section 3.3.1. Control of the parameters of an established and qualified process can provide more reliable assurance of an effective operation than test methods which may be inadequate.

1.6.1.   Cleaning Solution Properties

The surface tension, wetting, viscosity, solvency, and capillary action properties of the cleaning solution used are of prime importance to the success of a cleaning operation. A wetting index has been used as a measure of the capillary penetration potential of cleaning agents and is related to the density, viscosity, and surface tension.[16] Another factor employed to characterize solvents is the Kari butanol value, a relative measure of the ability to dissolve nonpolar compounds.[17] However there is no single scale by which cleaning solution power can be rated in absolute terms.

Wetting ability is inversely related to viscosity and surface tension. Capillary action (drawing liquids into small spaces) is directly related to surface tension and inversely related to density. High surface tension aids capillary penetration while low surface tension and low viscosity enhance circulation. Therefore, it is not always readily apparent what the effectivity of a particular solution will be in a specific cleaning assignment. It is necessary to consider factors other than the wetting properties or spreadability of a cleaning media on the surface to be cleaned. Wang and Seghal illustrate that "the penetrating power of the liquid will be directly proportional to the surface tension of the liquid and not inversely proportional...." Their results led to the conclusion "that under surface mounted devices, the predominant mechanism is capillary driven and it favors liquid with high surface energy."[18] Bonner in Chapter 3 and Munie in Chapter 4 of this book provide data for solvent and aqueous systems respectively.

1.6.2.   Process Enhancements

The mass transport capability of a cleaning process (the volume and flow rate or velocity of the solution able to contact the residues) is a major factor in determining the effectivity of the operation. With state of the art PWAs, the cleaning solution requires enhancements to increase mass transport. The process and equipment used to provide aid in contacting the residue sites and increasing mass transport include: increased temperature, vapor condensation, liquid sprays, ultrasonics, and rotation of the PWAs. Once the residues are solubilized or mobile and flushed off the assembly, it is necessary to prevent redeposition. Agitation addresses this need, but "purity" of the solution must be maintained with suitable recycling and cleaning.

Current major innovations in cleaning systems are directed toward additional methods of adding mechanical or thermal energy to the cleaning solution. Imparting energy appears to be more effective, potentially less detrimental to components and substrates, and generally a more viable approach for the long run than use of high power chemistry cleaning media.

Heating the cleaning solution is one source of adding energy to the system. The use of heated solutions will provide a decrease in viscosity and an increase in solubility of the contaminants encountered. Rosin fluxes significantly soften or "melt" at temperature in excess of 85 degrees C. Increased temperatures also result in improved penetration by the cleaning solutions.

Sprays have been used quite extensively in PWA cleaning and the trend is toward higher velocity applications through changes in orifices and pressures used. The effectiveness is dependent on the spray pattern, impact angle, distance of the nozzle from the PWA and the pressure-velocity-volume parameter relationship. Sprays are required for current technology cleaning systems to aid irrigation of close-packed assemblies.

Direct high pressure sprays are currently a more viable alternative than ultrasonic agitation. With an increase in the data base of experience with the use of ultrasonics as an aid in PWA cleaning, this may change. Use of sprays has been said to create electrostatic charges of a magnitude sufficient to damage highly sensitive components. One case in point involved radiation hardened devices unable to tolerate more than 30V of static.[19]

The use of ultrasonic energy to enhance the cleaning process is receiving increased attention as the density of PWA packaging increases. In late 1989 the evaluation of the technique was entering final phases at the Electronic Manufacturing Productivity Facility.[20] Use of ultrasonics in combination with sprays enhanced the cleaning effectivity achieved in a study reported by Elliot and Gileta.[21] An increase in the rate of softening of flux residues was observed when using ultrasonic energy rated at 3.9 watts/sq.inch at a frequency of 50 kHz in the immersion stage of a solvent cleaning system. For additional details of the status of advances in ultrasonic cleaning processes, refer to Johnson and Hayes in Chapter 5 of this book.

When using added mechanical energy from either high pressure/high velocity sprays, ultrasonics, or spinning the assemblies in the cleaning solution, consideration should be given to possible effects on very fine pitch (less than 0.025

inch) package solder joints. While the author knows of no cases where "good" joints have have failed as a result of high pressure sprays, it would seem advisable to consider the possibility during early process qualifications.

1.7. SUMMARY

As has been emphasized throughout this chapter, the cleaning solution and system selected for a specific application are affected by and affect many elements of the product and the total manufacturing process. In the technology of cleaning (as with many other things in our experience), "whenever you begin to describe something...whenever you try to pick out anything by itself you find it hitched to everything else in the universe."[22]

Effective postsolder cleaning of PWAs requires close attention to the intended use, design, and packaging of the product. The reliability requirements will impact the needs and extent of the cleaning operation. This must be factored into component selection and assembly manufacturing techniques with the realization that close-spaced, dense assemblies typical of today's product pose increased challenges to the cleaning process. To remove potentially harmful contaminants, there exists the necessity of forcing cleaning media through spaces with ever decreasing crossection.

The PWA manufacturing operations of assembly and soldering can introduce a myriad of contaminants. The cleaning media and process must be capable of solubilizing and removing both ionic and nonionic residues while mobilizing particulate contaminants, without damaging the product.

Current mixed technology PWA packaging incorporates multiple assembly and soldering cycles. It is expected that this will be the packaging technology of choice for most of the 1990s. Design and control of product and process are required in order to provide end products free of residue capable of degrading performance. The task is complex and must be accomplished within the constraints imposed by the need to preserve our environment.

The following chapters will address specific PWA cleaning/defluxing processes and materials which appear applicable at this time. Approaches described include the use of fluxes which leave limited residues which may be benign in selected service environments and indepth assessments of current and evolving solvent, aqueous, and alternative cleaning processes. Evolving packaging technologies in an era of valid concern for environmental issues will require continued innovation in all these areas.

REFERENCES

1. Tuck, J. "Cleaning Surface-Mounted Assemblies," *Circuits Manufacturing,* January 1984, p. 40.

2. Lermond, D. S. "Key Process Factors for Efficient Fluorosolvent Spray Cleaning of SMAs," *Printed Circuit Assembly,* May 1987, p. 22.
3. Musselman, R. P. "Shear Stress Cleaning of Printed Wiring Boards and Assemblies," Technical Program Proceedings, NEPCON East, Boston, MA, June 1985, p. 244.
4. Bonner, J. K. "Effective Cleaning of Surface Mount Assemblies," *Printed Circuit Assembly,* August 1988, p. 8.
5. Willis, R. "SMD Cleaning—A Practical Assessment," *Printed Circuit Assembly,* October 1987, p. 9.
6. Wang, A. E, and K. C. Seghal, "Effects of Wetting and Capillary Action on the Cleaning of SMAs," *Printed Circuit Assembly,* August 1988, p. 16
7. Cabelka, T. C., and W. Archer. "Critical Free Energy in Wetting: The Key to Maximum Effectiveness in SMA Cleanings," *Surface Mount Assembly,* December 1987, p. 20.
8. Paul, J. T. "Ultrasonic Cleaning—Can Current Technology Meet the SMT Challenges?", IPC-TP-798, October, 1989.
9. Gruenwald, F. "Aqueous Cleaning of Reflowed Surface Mount Assemblies," Technical Program Proceedings, NEPCON West, Anaheim, CA, February 1989, pp. 1024–1036.
10. Taylor, J. R. "Printed Board Assembly Cleanliness—Standards, Practice and Reliability Considerations," Seminar: Soldering in Electronic Production, Australian Tin Information Centre, August 1982.
11. Lermond, D. S. Op. cit., p. 22.
12. Wolff, M. "High Pressure Spray and Ultrasonic Cleaning for Surface Mount Assemblies," *SMT Forum,* December 1987, p. 14.
13. Elliott, D.A., and J. Gileta. "In-Line High Pressure Solvent Cleaning of Surface Mounted Assemblies—Part II," Technical Program Proceeding, NEPCON West, Anaheim, CA, February 1987, pp. 7.
14. Lermond, D. S. Op. cit., p 22.
15. Beer, F. "Cleanliness Testing of SM Assemblies," *Printed Circuit Assembly,* October 1989, pp. 48–51.
16. Kenyon, W. G. "New Ways to Select and Use Defluxing Solvents," Technical Program Proceedings NEPCON West, Anaheim, CA, February 1979.
17. Brous, J. and A. F. Schneider, "Cleaning Surface-Mounted Assemblies with Azeotropic Solvent Mixtures," *Electri-onics,* April 1984, p. 54.
18. Wang, A. E., and K. C. Seghal. Op. cit., p. 40.
19. Tuck, J. Op. cit. p. 40.
20. "Ultrasonic Cleaning for Military PWBs," EMPF TB 0008, U.S.Navy Electronic Manufacturing Productivity Facility, Ridgecrest, CA, Fall 1989.
21. Elliott, D. A., and J. Gileta. Op. cit., p. 6.
22. Muir, J. *My First Summer in the Sierra,* Chapter 6, p. 157 (paperback edition), Houghton Mifflin Co., 1916.

2
Flux Considerations with Emphasis on Low Solids

Leslie Adler Guth
AT&T Bell Laboratories

2.1. INTRODUCTION

2.1.1. Purpose and Chapter Description

To select cleaning materials, processes, and equipment without understanding the basic nature of the residues one wishes to remove is imprudent. In fact, the cleaning strategy is best chosen by taking into account the activity and the composition of the soldering flux used. Other variables associated with both the soldering process and the monitoring of the flux material are also important.

For these reasons, this chapter is devoted to soldering fluxes. This chapter begins with a section on the existing specifications for soldering fluxes and the associated test procedures for qualifying them. Following these, there are sections on flux composition, flux application methods, flux monitoring techniques, soldering process issues, all of which are directed toward wave soldering. The last three sections discuss nonliquid fluxes, soldering parameters, and future trends. Throughout this chapter, the characteristics of soldering fluxes and their use are described, with the underlying theme relating to the importance or the lack thereof for postsolder cleaning.

Fluxes are discussed in terms of chemical composition and activity so that the need for cleaning can be assessed. The four families of fluxes included are rosin, water soluble, synthetic activated, and low solids.

Due to the ever increasing environmental concerns, an emphasis is placed in this chapter on those fluxes and associated soldering processes that require no cleaning, especially low solids fluxes. Again, flux selection should be made realizing that this decision will directly affect the possible subsequent cleaning materials and processes and their associated environmental concerns.

2.1.2.   Definition of Soldering Flux

Soldering flux is defined in the flux specification of the Institute for Interconnecting and Packaging Electronic Circuits (IPC), ANSI/IPC-SF-818, "General Requirements for Electronic Soldering Fluxes," as:

> A chemically and physically active formula which promotes wetting of a metal surface by molten solder, by removing the oxide or other surface films from the base metals and the solder. The flux also protects the surfaces from reoxidation during soldering and alters the surface tension of the molten solder and the base metal. (IPC 1988, 3)

That is, flux prepares the metal surfaces to be soldered by:

- Removing oxides prior to soldering
- Assisting in uniform heating of the metal surfaces during pre-heat
- Maintaining an oxide-free surface during the soldering process
- Lowering the surface tension at the metal/solder interface

It is extremely difficult to solder without flux. Some processes have been developed recently that are supposedly fluxless; however, these processes typically use gaseous fluxes, and, in some cases, spray liquids that are not commercially available fluxes, but are composed of the materials common to commercial fluxes (e.g., a mixture of alcohol and an organic acid). In other words, materials that meet the definition of "flux" are certainly used in these processes as well, even though these materials may not necessarily be marketed and commercially available as "fluxes."

So the vast majority of soldering is done with the aid of fluxes; choosing the appropriate flux for the soldering process dictates the necessary postsoldering cleaning process.

2.2.   SPECIFICATIONS

All flux specifications are linked by a common thread: descriptions of test methods and requirements that must be met; however, they all differ by the specific tests and requirements that are imposed. In fact, they even differ by the ways that fluxes are categorized or in the general nature of what types of fluxes are allowed.

Many specifications classify fluxes in terms of material content such as:

1. Rosin, water soluble, synthetic activated, or low solids
2. Rosin, rosin mildly activated, rosin activated, or rosin superactivated

However, other specifications group fluxes in terms of their activity, corrosion potential of residues, and the need for rigorous cleaning such as:

1. Low activity, medium activity, or high activity (e.g., IPC SF-818)
2. Compliant, or noncompliant (e.g., Bellcore TR-TSY-000078, Issue 2)

Before examining the different specifications in detail, several test methods which are cited in two or more specifications will be discussed in a general sense. These methods are: the copper mirror test, the silver chromate paper (halide) test, the surface insulation resistance test, and the British corrosion test.

2.2.1. Test Methods

Each of the methods discussed below have specific purposes, advantages, and disadvantages. These will be described along with their procedures.

Copper Mirror Test. The copper mirror test is referenced in a number of specifications, and is a measure of a flux's corrosivity to copper. This corrosivity is an indication of the activity level of a flux. Both the speed and relative ease with which this test can be performed make its use quite common. However, it also has several drawbacks.

The copper mirror test involves placing two drops of the flux in question on a copper mirror. Although the IPC test method IPC-TM-650 Sect. 2.3.32 calls for one drop only, it is clear from the examples included in the test method that two drops are used (common practice is the use of two drops simply to increase confidence). A copper mirror is a glass microscope slide on which has been vapor deposited approximately 300–500 Angstroms of copper, defined as a Cu thickness which allows $10 \pm 5\%$ transmission of normal incident 5000 Angstrom light. Typically a flux made of pure water white gum rosin and isopropanol acts as a control and a drop of it is also placed on the mirror. The mirror is then placed in a controlled environment of $23 \pm 2°C$ and $50 \pm 5\%$ relative humidity for 24 hours. Then the mirror is rinsed in isopropanol and examined for any spots or areas where the copper has been removed by the flux. Examples of fluxes that passed and failed the copper mirror test are shown in Figure 2.1.

Results of the copper mirror test indicate the potential corrosivity of the raw or unheated flux. Although the corrosivity of unheated flux can be a concern, it is unfortunately not always indicative of the corrosivity of a partially heated flux (e.g., flux present on the top or component side of a circuit pack during wave soldering) or a fully heated flux (e.g., flux present on the bottom or wiring side of a circuit pack during wave soldering). Also, the results can be difficult to interpret sometimes since it is can be somewhat subjective and there is no middle ground between pass or fail.

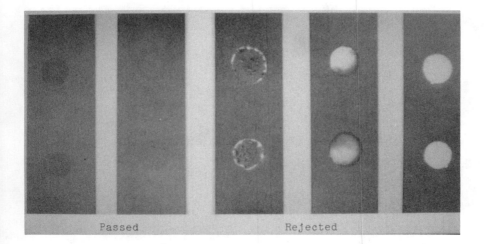

Fig. 2.1.   Results of copper mirror test.

Because of these drawbacks, critics have questioned the relevance of this test altogether. However, due to the speed, the relative ease of performing this test, and the evidence that these results do correlate well with those of the more lengthy and involved surface insulation resistance test (Guth 1988), this test is certainly worthwhile and justified. It is especially useful as a pre-screening test when testing a new formulation or as a routine incoming materials check.

Silver Chromate Paper (Halide) Test.   The silver chromate paper test is another fairly simple test to perform. Its purpose is to detect significant amounts of chlorides and bromides in a flux. Any flux that leaves the silver chromate paper unchanged has either no Cl^- or Br^-, or at least no detectable quantities.

The test procedure involves placing a drop of flux on silver chromate paper. After 15 seconds, the paper is rinsed or immersed in isopropanol. The paper is allowed to dry for 10 minutes and then is examined for any color change. A color change means that halides are present in detectable quantities. Examples of silver chromate paper results are shown in Figure 2.2.

Drawbacks of this test include possible subjectivity in interpretation of results, lack of quantitative data, and interference from other nonhalide materials. Some chemicals can cause color change, as can fluxes that are unusually acidic (pH < 3). Thus, if halides are not suspected and a color change is detected, other tests should be performed to verify the results.

Another problem with this test is that the results are pass/fail only, and are not quantitative. A proposal in a telecommunications specification had called for a more quantitative assessment of the amount of chlorides and bromides in

Fig. 2.2. Results of silver chomate paper test.

a flux based on a color changes seen in the silver chromate paper (Bellcore 1987). Yet another problem is that the silver chromate paper only detects chlorides and bromides, not fluorides.

Even with these shortcomings, though, this test is beneficial for prescreening or incoming material inspection for fluxes that should be noncorrosive.

Surface Insulation Resistance Test. Surface insulation resistance (SIR), also known as insulation resistance (IR) or moisture and insulation resistance (M&IR), is a lengthy but worthwhile flux evaluation test. Of all test methods, this one best simulates the relevant cause and effect relationship between soldering fluxes and circuit boards. This is true because: (1) the soldering process is part of the sample preparation, (2) the sample substrate is itself a circuit board, and (3) the response variable, leakage current, is a measured attribute of all circuitry. Even so, this test method does have some shortcomings which will be discussed along with the details about the types of test patterns, sample preparation, temperature and humidity selection, necessary equipment, and types of requirements.

The result of SIR testing is an electrical measurement of the effect of flux and/or cleaning residues on a printed circuit board. To maximize the measuring sensitivity of this test, the leakage current passing between long parallel lines placed at different electrical potentials is measured. Instead of long narrow circuits, interdigitated combs which are electrically identical to two long lines are typically used. Examples of two comb patterns are shown in Figure 2.3, (a) the IPC B comb pattern with 0.0125″ lines and spacings and (b) the telecommunications comb pattern with 0.025″ lines and 0.050″ spacings.

Traditionally, data are reported as insulation resistance (in ohms or megohms) rather than as leakage current. This conversion is simple via Ohm's Law, $V = IR$, where V is the test voltage and I is the leakage current. Diagrams in Figure 2.4 show the typical circuitry used to apply a 45–50 VDC bias voltage on the left and to measure the resistance with a reverse polarity 100 V test voltage on the right. SIR data are a function of a number of parameters:

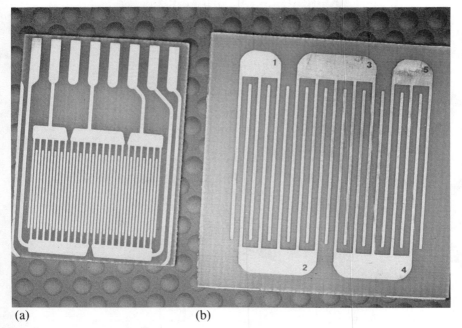

Fig. 2.3.   Two surface insulation resistance comb patterns: telecommunications and IPC.
(a) IPC Comb Pattern (b) Telecommunications Comb Pattern

1. The width of the spacing between the traces
2. The length of the traces
3. The testing voltage
4. The test voltage electrification time
5. The bias voltage, if any
6. The temperature
7. The relative humidity
8. The dielectric material of the test substrate
9. The specific solder resist coating, if any

Although SIR is affected by both the total length and width between the traces, it is a moot point when testing fluxes if the same pattern is always used. However, there are many comb pattern shapes and sizes. Theoretically, the data from all comb patterns can be correlated by taking into account the number of squares that make up a comb. A square is a unitless value equivalent to the length of the comb divided by the width of the spacing between two lines. So, a comb pattern that has more lines will be longer and have more squares; additionally, a comb pattern that has narrower spacing will have more squares. Several papers

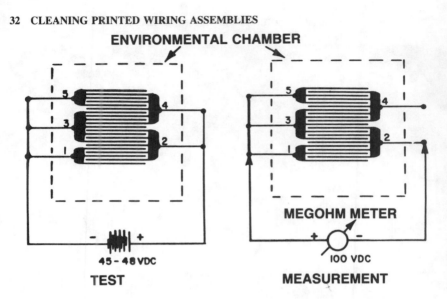

Fig. 2.4.   Circuit diagrams for biasing and measuring SIR comb patterns.

which have discussed theory and concepts behind the use of squares are recommended for further reading (Zado 1984; Gorondy 1984). More recently, a
study has been documented showing that correlation is good within a range of
squares of not more than one order of magnitude (Chan and Shankhoff 1989).

The test voltage as well as the electrification time can also influence the ultimate
SIR measured. Moreover, the bias voltage and its value relative to the test voltage
can affect the results. These results have been documented (Gorondy 1988).
Whether to use a bias voltage and whether this bias voltage should be the reverse
polarity of the test voltage is an ongoing debate. Both the telecommunications
and the IPC specifications require a 100 VDC test voltage and a 45–50 VDC
bias of reverse polarity. Although these electrical parameters may influence the
SIR results, it is important to look at them in the context of the whole picture.
That is, if one simply maintains particular settings for these parameters, then
the changes in the SIR resulting from changes in the electrical parameters can
be disregarded. Established SIR test methods specify these settings so the relative
differences between different fluxes can be observed without worrying about
electrical parameter effects.

Both temperature and humidity can be used to accelerate the aging of samples.
Depending on the settings, however, these factors can also be specified to simulate worst case conditions. Figure 2.5 shows several typical temperature and
humidity settings. Also shown is the envelope in which circuitry typically must
operate relative to an example of a performance test condition (e.g., 35°C at
90% RH) and several life time or accelerated test conditions (e.g., 55°C, 65°C,

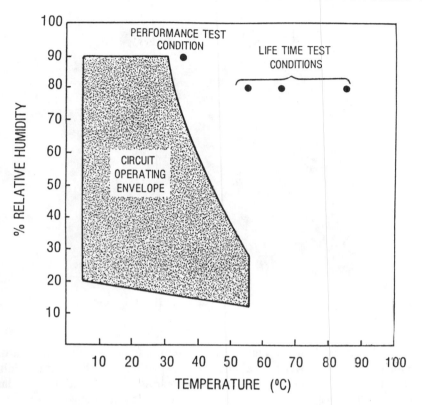

Fig. 2.5.   Typical temperature and humidity conditions for SIR tests.

85°C, all at 80% RH). As desirable as accelerated conditions may seem, it is important to realize that an extreme environment may cause some flux materials to fail by mechanisms that are nonexistent at normal operating conditions.

Sample preparation simulates the assembly process, including fluxing the comb pattern and then soldering it. Recently, the Bellcore specification has added an additional test in which component side flux residues are tested by floating the comb pattern face up on a solder pot or wave, rather than face down. This additional test is important since some flux residues can potentially be more harmful when exposed to less heat.

Equipment needed to perform SIR testing includes electrical measuring instruments and environmental chambers. Obviously, the more automated the setup the better. A good SIR setup allows for ease in testing many samples simultaneously, requiring some sort of switching mechanism or matrix box to individually test all samples. An example of commercially available equipment which can be programmed to apply a bias, measure the SIR at regular intervals, and

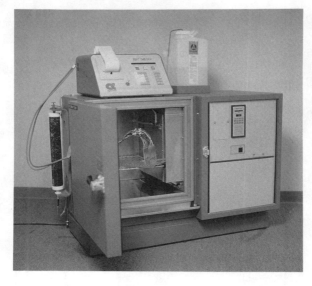

Fig. 2.6. Commercially available data acquisition system for SIR tests. (Courtesy of Alpha Metals.)

record SIR data is shown in Figure 2.6. For temperature and humidity control, one may use either an oven with a saturated salt solution controlling the humidity or a chamber with a microprocessor which controls both temperature and humidity. The latter is preferred simply because there is less maintenance compared to checking a saturated salt solution.

Some specifications require minimum absolute SIR values as a pass/fail criterion, whereas others require a minimum SIR relative to the SIR value of a control. A relative value is preferred since it is simply allowing a minimum degradation due to the presence of flux residues. In contrast, an absolute value for an SIR requirement implies a knowledge of the electrical needs for a particular product, a difficult if not impossible task for widely used industry specifications.

British Corrosion Test. Another test that is specified by a number of groups is a corrosion test, also referred to as the British corrosion test. It involves reflowing solder in the dimple of a copper panel in the presence of flux solids, and then aging the panel at 40°C and 93% RH for a number of days, depending on the specification and the type of flux being tested. A flux is said to pass if no blue-green corrosion can be observed after aging. Figure 2.7 shows both passing and failing results for the corrosion test.

This test has both benefits and drawbacks. On the plus side, the results are independent of a flux's solids content, since this quantity is held constant. Be-

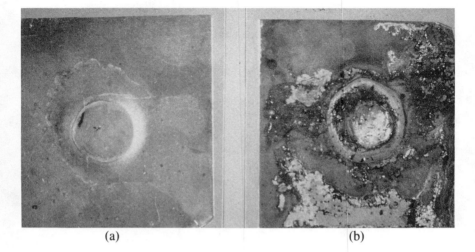

(a)   (b)

Fig. 2.7.   Results of corrosion test. (a) Passing (b) Failing

cause a concentrated flux material is used, the test is certainly a worst case condition.

However, the test lacks an electrical bias, which is necessary to initiate some corrosion mechanisms (e.g., electrolytic). Also, it creates an accentuated cooper/solder/flux interface. Lastly, the interpretation of results is difficult, subjective, and qualitative only. An evaluation of "how green is green" and the extent of green or blue-green coloration is difficult to quantify. As with the copper mirror test, there is no middle ground between pass and fail.

The difference between this test and the SIR test lies in their distinct purposes. The corrosion test evaluates the potential for an "open" failure, whereas the SIR test determines the potential for an electrical "short."

2.2.2.   IPC

The IPC flux specification, IPC-SF-818 classifies fluxes in terms of activity levels, and corrosivity potentials of their residues rather than in terms of material types. All electronic grade fluxes should fit in one of three major categories of L, M, and H, denoting low, medium, and high activity, respectively. Because of the further groupings for the SIR testing, ultimately a flux would be categorized with L, M, or H, followed by the numeral 1, 2, or 3 (designating class of assembly) and by C, N, or CN depending on whether a flux meets the SIR after being cleaned (C), not cleaned (N), or both (CN).

Because of the classification system itself, virtually all fluxes can fit into at least one category. It is up to the user to determine what product class is being soldered and whether postsolder cleaning is part of the assembly process. After these decisions, the activity level necessary or sufficient for the assembly process must be chosen, remembering that with higher activity level, there is usually increased soldering efficiency along with increased potential for corrosion if proper cleaning procedures are not maintained.

The IPC specification requires: (1) copper mirror test, (2) silver chromate (halide) test, (3) corrosion test, and (4) surface insulation resistance test. The SIR comb pattern is part of the IPC-B-25 test board and has 0.0125 inch lines and 0.0125 inch spacings (see Figure 2.3). The SIR requirement is an absolute value of $1 \times 10^8 \ \Omega$.

2.2.3. U.S. Military

The U.S. Military specification, MIL-F-14256E, is limited to only rosin-based fluxes. There are three categories allowed, R, RMA, and RA designating rosin, mildly activated rosin, and activated rosin, respectively. These are equivalent to the R, RMA, and RA categories in QQ-S-571, the U.S. government's solder specification.

Like many other specifications, MIL-F-14256 requires a copper mirror test (for R and RMA) and a silver chromate test (for R, RMA, and RA). A solder spread factor is required for RMA and RA fluxes.

However, what sets the military specification apart from other specifications is that it not only allows just rosin fluxes, but also demands that a minimum of 51% of the solids content is rosin, and that the rosin itself must meet a certain minimum acid number. Another unusual requirement is a minimum resistivity of water extract. This test involves measuring the conductivity of a beaker of distilled water, then adding a small quantity of flux, boiling the mixture for 1 minute, cooling it, and then measuring the conductivity of the cooled liquid. There is a minimum allowed resistivity for both the uncontaminated distilled water and the final solution, the latter requirement depending on whether the flux is an RMA or RA. The relevance of this test has been debated over the years and is still ongoing. It is measuring the conductivity of water soluble and ionizable species in the flux which may or may not be corrosive either after being exposed to soldering temperatures or in the presence of other nonconductive encapsulating materials.

For the first time, MIL-F-14256 has incorporated a surface insulation resistance test in Issue E dated June 1989. It requires a 100 MΩ reading under 85°C and 85% RH conditions using the IPC B-25 B comb pattern.

2.2.4. Telecommunications

The telecommunications industry has its own set of specifications and requirements. After the divestiture of AT&T in 1984, the court set up a research group to work for the interests of the seven Bell operating companies, Bellcore. The Bellcore specifications grew out of the original AT&T specifications, and have in recent years, dictated telecommunication industry requirements.

The most recent Bellcore assembly document which specifies flux materials is TR-TSY-000078, Issue 2 (Bellcore 1988). It divides allowed fluxes into only two categories, compliant and noncompliant.

Compliant fluxes are "noncorrosive fluxes" (formerly called "rosin noncorrosive fluxes"). Requirements which must be met include a copper mirror test, a silver chromate test, a fluoride test, a pH test, and a surface insulation resistance test. The SIR test specifies a comb pattern with 0.025 inch lines and 0.050 inch spacings (see Figure 2.3) and an environment of 35°C and 90% RH. Originally, Bellcore required that a comb pattern soldered with a compliant flux be at least 75% of the SIR of the control (like the AT&T specification MS-58556). Even though Bellcore still requires an SIR value relative to the control, the requirement is really the equivalent resistance of two parallel resistors, one being the SIR for the control and the other an ideal resistor with a value of 10^5 MΩ. Thus, the requirement is a sliding scale and the gap between it and the control narrows as the control value decreases. This requirement must be met for three different types of comb pattern groups: (1) fluxed, soldered, and cleaned, (2) fluxed, soldered, and not cleaned, and (3) fluxed, soldered with comb face up, and not cleaned. The reason for the third group is to simulate the exposure of the top or component side of a circuit board to flux. This is necessary because some fluxes have been found to be more detrimental when they are not heated to as high a temperature, a situation experienced by flux present on the top side of a circuit board.

There is a misconception that a telecommunications noncorrosive rosin or compliant rosin flux is the same as an RMA flux. This is not necessarily true, since the requirements for each are different. For example, just because a flux meets the military's water extract resistivity requirement does not mean that the same flux will meet the telecommunications SIR requirement. In fact, a number of fluxes meet one of these specifications and not the other.

The other category for telecommunication fluxes is noncompliant, which is simply a classification for aggressive fluxes. That is, any flux that cannot meet the requirements for compliant fluxes is then tested for the noncompliant requirements. Aggressive fluxes are tested per these two requirements: (1) surface insulation resistance of 3×10^9 Ω on a telecommunications comb pattern with stripes of solder mask perpendicular to the comb's metal traces exposed to the flux, solder, and an appropriate cleaning process, and (2) ionic extraction value of 1.0 $\mu g/cm^2$ (6.45 $\mu g/in.^2$) NaCl equivalent for a bare copper comb pattern exposed to flux, solder, and the cleaning process. (For a description of the ionic extraction test, see Chapter 3.) Clearly, not all aggressive fluxes can necessarily meet these requirements. Since the residues of these more active fluxes are designed to be removed by cleaning, testing uncleaned boards is not required.

The inclusion of solder mask stripes on the SIR coupon has been debated as well as possible alternative patterns. Some people fear that the solder mask stripes make the test vehicle unduly stringent and even unrealistic compared to actual circuit designs. Effort to modify this geometry is ongoing. Unlike the SIR test, however, the ionic extraction is well accepted by industry as a cleaning efficiency check when using aggressive fluxes. The ratio of alcohol to water used for the extraction medium can be either 50 : 50 or 75 : 25.

2.2.5.   Summary

There are almost as many ways of categorizing fluxes as there are flux specifications. Some only allow rosin based fluxes, others categorize fluxes on a material basis, while still others group by flux and flux residue activity. A number of tests are pervasive throughout the industry including copper mirror and surface insulation resistance. The copper mirror test is simple to perform, while the SIR test is an electrical test, best simulating the effect of a flux on an actual product. Consequently, when selecting a flux and the appropriate specification to which it should be tested, one should consider the ultimate use of a product and the customer's material requirements.

2.3.   FLUX MATERIALS

Fluxes are composed of a number of different materials including:

- Solvent. The liquid carrier for the flux ingredients, allowing even distribution of the flux material on the printed circuit board. During the preheating of the board, the solvent is intended to evaporate, so little, if any, is present when the board contacts the solder wave. For most fluxes, the solvent is isopropanol and/or some other alcohol.
- Vehicle. A thermally stable material that acts as a high temperature solvent during wave soldering. Sometimes it is also a weak activator (e.g., rosin). Depending on its properties, it can inhibit or induce corrosion. Specific vehicles will be discussed further in the following sections on flux types.
- Activator. One or more ingredients in a flux that create a wettable surface for solder by removing oxides and possibly other contaminants when coming into contact with the metal on the printed circuit board. These ingredients may or may not be corrosive at room temperature, but certainly are active

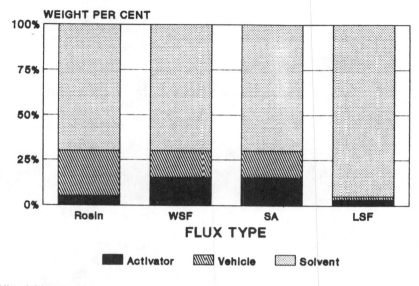

WEIGHT PER CENT

FLUX TYPE

■ Activator   ▨ Vehicle   ▦ Solvent

All weights approx.

Fig. 2.8. Comparison of major component proportions for four groups of fluxes.

or corrosive at elevated temperatures to perform their job properly. Activators are typically organic solids and organohalides.

- Surfactant. An ingredient that reduces the surface tension at the metal/solder interface to further promote solder wetting, especially when the printed circuit board exits the solder wave. It can also serve as a foaming agent, sometimes a necessary additive so that a flux can be applied with a foam fluxer.
- Antioxidant. A material that prevents reoxidation of the metal surfaces after the activator has prepared them for soldering. Often, another material such as the vehicle might also serve as an antioxidant.

In this section, flux characteristics will be discussed in greater detail, and will be examined in terms of the following flux categories: rosin, water soluble, synthetic activated, and low solids.

Before looking at each flux group individually, it is worthwhile to compare the makeup of these fluxes in terms of generic constituents and proportions, only looking at the three major components, solvent, vehicle, and activator. As shown graphically in Figure 2.8, these four groups of fluxes consist of different proportions of vehicle, activator, and solvent. Those rosin fluxes which pass military RMA or telecommunication specifications typically have a much lower activator to vehicle ratio than the other flux types shown. Water soluble and synthetic

CLEANING MEDIUM	FLUX TYPE				
	Rosin (benign)	Rosin (aggressive)	WSF	SA	LSF
None (no clean)	X	-	-	-	X
Chlorinated solvents or CFC-113 based	X	X	-	X	-
Aqueous (saponifier)	X	X	X	-	-
Semi-aqueous (terpenes)	X	X	X	X	-

Fig. 2.9.   Cleaning material choices as a function of flux type.

activated fluxes may have about the same proportion of activator relative to the amount of vehicle. All of these groups typically have 15 wt. % to 35 wt. % solids or nonvolatiles. In comparison, low solids fluxes have much less solids, and the amount of activator is typically the same or more than the amount of vehicle.

Figure 2.9 shows the effect of flux selection on cleaning material choice. Not all cleaning materials can remove all flux residue types, nor do all flux residues need to be removed.

2.3.1.   Rosin

The vehicle in rosin fluxes is rosin or colophony, a naturally derived material from pine trees. In the United States, the word *rosin* usually refers to the natural substance whereas the word *resin* refers to a synthetically made material. In the UK, the word *resin* is used instead of *rosin*. The rosin material normally used in fluxes is water white rosin, a high grade material, similar to compounds used in varnishes and lacquers.

Rosin material has several unique characteristics. Besides acting as a flux vehicle, it is also a mild activator at soldering temperatures. Rosin consists of a number of isomers, mostly abietic acid, which are fairly large and include one carboxyl group (molecular formula: $C_{19}H_{29}COOH$). Another unique property of rosin is that it is an extremely good insulator at room temperature; in fact, its bulk resistivity is at least an order of magnitude higher than the epoxy-glass printed wiring board material's bulk resistivity (Zado 1983a). Because rosin hardens on cooling to room temperature, it serves as an excellent encapsulant, prohibiting the movement of other flux residue ingredients such as those from activators. However, as mentioned already, at soldering temperatures rosin is mildly active, so that, dissolved in an appropriate solvent such as isopropanol, it can perform all the necessary functions of a flux. Usually, though, the activity of rosin by itself is not sufficient for production soldering most components and printed wiring boards.

Depending on the final activity of a rosin flux, a number of different types of activators are typically used, ranging from organic acids to organohalides. Again, depending on the quantities added, rosin fluxes may or may not pass copper mirror, silver chromate, surface insulation resistance, and corrosion tests. L or RMA fluxes should pass these tests, whereas RA, M, or H will not.

In the telecommunications industry, rosin fluxes are tested per the compliant flux test requirements. The U.S. military only allows rosin based fluxes; however, as discussed earlier, the group of rosin fluxes allowed by telecommunication specifications overlaps somewhat, but not totally, the group allowed by the military.

It is important to note that rosin will undergo a change around 75°C to 85°C, its softening point, which can impact accelerated aging or surface insulation resistance tests performed at or above this temperature range. By softening, the rosin loses its encapsulating properties, allowing usually tightly bound flux ingredients to be more mobile. Sometimes, rosin flux residue has been erroneously categorized as harmful because it has been tested above the softening point; that is, if a printed circuit board will never see operating conditions this high, then a failure mode observed during high temperature aging conditions is one that will never actually be seen in the field, and is thus irrelevant.

Work has been carried out over the year to maintain the noncorrosive nature of rosin fluxes, yet maximize the soldering efficiency achievable with them. Additional reading on rosin flux development is available (Zado 1983b).

Because of the insulating properties of rosin flux residue, it is not necessarily cleaned off to avoid corrosion. In fact, many companies have left rosin flux residue unremoved for years without problems. Nevertheless, there are a number of reasons for removing rosin flux residue. Often rosin residue is sticky and tacky. So, for aesthetic reasons or if the printed circuit boards may be exposed to excessive dust and dirt (Reagor and Russell 1986), it may be necessary to clean. Another reason for cleaning is to avoid problems during bed-of-nails testing or in-circuit testing; rosin flux residue can cause false opens when the test pins are unable to push through insulating residue covering a test pad. Sometimes it is necessary to remove rosin residue because of its effect on the subsequent addition of conformal coating to a printed circuit board. It is unfor-

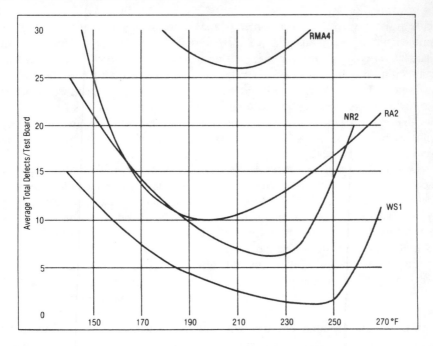

Fig. 2.10. Solder defect levels as a function of preheat temperature for rosin fluxes RA2, NR2, and RMA4 compared to water soluble flux WS1 (Chung, Cassidy, and Graham, 1983).

tunate when it is necessary to remove rosin residue, since its complete removal, although possible, is often quite difficult.

Cleaning alternatives and the associated cleaning equipment for removal of rosin flux residue are discussed in a number of other chapters in this book. Alternatives include CFCs, HCFCs, aqueous detergents and saponifiers, and semi-aqueous solutions (terpenes).

2.3.2. Water Soluble

Water soluble fluxes (WSFs) are typically very active and aggressive, and the residue can be corrosive if not removed properly. However, since these fluxes are active and their residues are designed to removed in aqueous solutions, they offer a number of advantages over less active and more difficult to remove fluxes (Chung, Cassidy, and Graham 1983). In fact, with their use the process window for soldering is relatively large, and extremely low solder defect levels (less than 50 ppm) are attainable. Figure 2.10 illustrates the wider process window and the lower solder defect level attainable with a WSF compared to some rosin fluxes. With the improved soldering efficiency, some companies have been able to eliminate the need for post-solder touch-up operations.

Often WSFs are referred to as organic acid or OA fluxes. Although it is true that there are organic acids in many WSFs, it is also true that other fluxes including rosin, SA, and low solids include organic acids. Thus, the name OA for water soluble fluxes is a misnomer, and should be avoided.

The vehicle for WSFs is typically a polyethyleneoxide or polypropyleneoxide, commonly referred to as polyglycols. These materials are nonionic but hygroscopic, so a synergistic deleterious effect is observed when these residues are not removed along with the ionic residues from WSFs (Zado 1983a). That is, polyglycols do not encapsulate activator residues like rosin.

Usually the activity level is quite high in WSFs, meaning that fairly strong organic acids and halide bearing compounds are present or that relatively large quantities are present. Some WSFs are halide-free and promoted as less corrosive, yet their postsolder residue still needs to be removed because of the organic acid present.

Because of their high activity level, most WSFs will fail the copper mirror and silver chromate paper tests. If proper cleaning is employed, a number of WSFs can pass SIR and corrosion tests. Both WSFs and SA (see next section) fluxes must meet the requirements of noncompliant fluxes for the telecommunications industry, and neither are presently allowed by the U.S. Military since they do not contain rosin.

WSFs, like other aggressive fluxes, must be removed in a spray impingement or other energy enhanced cleaning system. For WSFs, the cleaning medium is either an aqueous only or an aqueous, detergent saponifier solution. Inline spray systems look much like car washes, with spray nozzles directed at the printed circuit boards from both the top and the bottom. Because of the need to thoroughly remove all residues and because of the type of cleaning used, the compatibility of components with immersion in water must be addressed. Also, it is important to ensure that components are irrigable. The use of WSF and aqueous cleaning for surface mount or mixed assemblies has been debated for a number of years. However, a number of companies have demonstrated their feasibility (Wargotz, Guth, and Stroud 1987; Danford and Gallagher 1987; Aspandiar, Piyarali, and Prasad 1986; Dickinson, Guth, and Wenger 1989). For more details about the WSFs and the type of cleaning necessary, see Chapter 4 of this book.

2.3.3.   Synthetic Activated

In the 1980s, Du Pont developed a family of high activity synthetic activated (SA) fluxes, analogous to WSFs designed so that the residues were easily soluble in the CFC-113/methanol azeotrope cleaning medium. SA fluxes are based on a mono- and diisooctyl phosphate (IOP) chemistry and include compatible activator and solvent materials (Kenyon 1986). This technology was offered to flux manufacturers who subsequently marketed these SA fluxes.

Because of their high activity, these fluxes perform much like WSFs in that they will not pass the copper mirror or silver chromate paper tests. Also, SA flux residues must be removed in a spray impingement or total immersion cleaning system. Usually the cleaning medium is chlorofluorocarbon (CFC) based; however, it has been shown that terpenes can also adequately remove SA flux residues (Hayes and Hood 1989). Thus, component compatibility with the cleaning medium is an issue that must be addressed. Unfortunately, a component that is compatible with water may not be compatible with CFCs, and vice versa. If the cleaning process is efficient, the requirements for SIR and corrosion tests can be met with SA fluxes.

Like WSFs, SA fluxes typically yield lower solder defects than rosin fluxes and their residues are more easily removable. Nevertheless, even with these benefits, due to environmental concerns with CFCs, a decline in SA flux usage is already being seen and should continue.

2.3.4. Low Solids (No-Clean)

Unlike the fluxes discussed in the previous sections, low solids fluxes (LSF) are formulated to leave minimal or no postsolder residue. As the name implies, this category of fluxes includes those that have a much lower solids content compared to traditional fluxes. In fact, the solids or nonvolatiles content ranges from 2 to 5 wt.%.

Because the intent is to avoid cleaning, these low solids fluxes should meet the LN (preferably) or at least MN activity designations specified in IPC-SF-818. They also should be able to meet the noncorrosive or compliant flux requirements in the telecommunication specification. LSFs are not presently allowed by the U.S. Military.

Benefits of Low Solids Fluxes. The elimination of postsolder cleaning eliminates machine cost and maintenance, and reduces materials and operating costs. Along with cost reductions, space savings and process simplification are also realized. Additionally, the need for any postsolder assembly steps is eliminated in cases where components, incompatible with the cleaning process, must be manually inserted and soldered. (This post-solder assembly of components is often necessary when aggressive fluxes like WSF and SA which require immersion cleaning are used.) Moreover, the associated temporary solder mask used as a solder resist for plated through holes for postsolder assembly is not needed.

Concerns over EPA regulations, international restrictions, component qualification, and waste disposal are precluded by the elimination of cleaning. The environmental concerns are becoming increasingly important in making process decisions as more is discovered about the effect of these processes on the ecosystems and atmosphere.

As unexpected benefit has sometimes been realized in terms of improved soldering yields with LSFs relative to those with rosin fluxes (Sholley 1989). In fact, as much as an order of magnitude drop in defect levels has been observed (Guth 1989a). These defect level changes are also dependent on the product code and process parameters. Because of the lighter consistency and freer flow compared to rosin fluxes, low solids fluxes can more easily wet all surfaces of the wiring side of a board, especially important when surface mount devices are wave soldered (Toubin 1989).

Characteristics of Low Solids Fluxes. Ideally, a flux that does not need postsolder cleaning should:

• Be noncorrosive
• Leave a nontacky, noncorrosive, colorless, and/or minimal amount of residue
• Have sufficient activity to yield acceptable soldering
• Be compatible with existing fluxing equipment

Unfortunately, it is difficult, if not impossible, to find an LSF that meets all of these characteristics. Instead, it might be necessary to modify the flux application process to realize all of the above.

When a low solids content is maintained and sufficient flux activity insured, the activator often outweighs the vehicle. Because of this unusual ratio, one cannot rely on the vehicle to encapsulate activator residues. For the more active LSFs that allow adequate soldering, this is the case. On the other hand, some flux manufacturers have made LSFs which are simply diluted versions of traditional noncorrosive rosin fluxes; the amount of rosin far outweighs the amount of activator. Experience with these fluxes, though, has shown that the activity level is not adequate for many soldering operations.

The solvent used in LSFs is typically isopropanol, the same one used in the other fluxes, the only difference being that there is significantly more solvent in an LSF that in a traditional solids flux.

The vehicle in LSFs is a synthetic resin material, rosin, or modified rosin. Those that contain rosin are amber in color, rather than clear, and can, like higher solids content rosin fluxes, leave residues that impair bed-of-nails testing. One flux manufacturer developed a number of LSFs based on a vehicle which is more thermally stable than rosin, pentaerythritol tetrabenzoate (PETB) (Rubin 1982). Besides being the flux vehicle, PETB supposedly acts as an antioxidant also.

Most LSFs are halide free and contain only organic acid type activators.

Usually, the ones that have halides also include rosin. Preferably, an LSF will be halide and rosin free.

Many LSFs do not naturally foam because of their low solids content and the characteristics of the particular constituents themselves. LSFs that are foamable usually contain foaming agents, which are either nonvolatile or volatile. Some nonvolatile foaming agents are glycols that remain along with the other flux residue after soldering. Like the glycols in WSFs, these can be hygroscopic, creating an environment for potential corrosion. Some LSFs include, along with isopropanol, small quantities of other solvents that promote foaming but volatilize during heating. In general, it has been seen that foaming agents may have a deleterious effect on copper mirror, SIR, and corrosion test results.

Results of Published Studies on LSFs. A number of publications about LSFs have discussed the process parameters, and application and flux monitoring difficulties that are specific to this flux class. These will be discussed in later sections dealing with these particular subjects.

It has been noted that LSFs may be relatively difficult to remove with traditional CFCs; in fact, white residues can form so caution is necessary (Toubin 1989). Others have also observed this problem. Consequently, it should be emphasized that there are certainly some cases when cleaning does not necessarily improve a situation, and can, in fact, do harm.

It has been shown that the surface insulation resistance decreases as a function of the original quantity of flux applied, inferring that excessive postsolder residue can cause electrical problems (Guth 1989a). The details of this work are described in the rest of this section.

In addition to the typical flux evaluation tests, simultaneous accelerated aging and functional testing of actual product is considered to be a critical step in implementing new assembly processes in AT&T. Accelerated aging of circuits soldered with LSFs resulted in failure of three out of eight circuit packs submitted to 65°C/85% RH for 250 hours with some components showing severe corrosion. Also, a white film on the dielectric was visible. The test was repeated with new circuit packs that were exposed to less LSF, and no failures were observed.

These results indicated that excessive low solids flux residues are detrimental to circuit integrity. In hindsight, it seems obvious that flux residues that are not removed after soldering and do not contain a large proportion of highly insulating water white rosin could cause problems. To better understand this phenomenon and confirm this hypothesis, SIR studies were modified such that the quantity of flux applied to comb patterns was carefully controlled.

The SIR testing described was performed on comb patterns with 25 mil lines and 50 mil spacings exposed to 35°C and 90% RH and 50 VDC bias using a 100 VDC reverse polarity test voltage. Since fluxes are actually nonvolatiles

CALCULATION FOR
FLUX VOLUME → WT. of FLUX APPLIED

$$\frac{W_{final} - W_{initial}}{A} = \frac{FNV \times SG \times V_{flux}}{A}$$

Where:

W_{final}	=	weight of substrate after flux application and drying
$W_{initial}$	=	weight of substrate before flux application
A	=	area of substrate
FNV	=	weight % of flux non-volatiles
SG	=	specific gravity of flux
V_{flux}	=	amount of flux applied by volume

Fig. 2.11   Relationship between the volume of liquid flux and the weight of the dried flux (Guth 1989a).

dissolved in alcohol and the alcohol readily evaporates at ambient conditions, the method of flux quantity control was based on volumetric measurements.

To monitor this application method, a correlation was made between the volume for the liquid flux to the weight of the nonvolatile flux materials. This relationship is shown in Figure 2.11, and is obviously flux dependent. Monitoring the flux application simply involves a weight measurement. A substrate is weighed, fluxed, allowed to dry, and weighed again. The amount of flux in terms of nonvolatiles per unit area of substrate is calculated. As with other SIR tests, after flux application, the comb patterns were wave soldered.

The SIR data for various quantities of a particular flux, Flux A, along with the SIR for control coupons are shown in Figure 2.12. The quantities of Flux A applied to the comb patterns are described relative to each other, where the lowest quantity is designated X, the next quantity used was twice X (2X), then five times X (5X), and finally, forty times X (40X).

As the flux quantity increases, the SIR decreases. Leakage paths were observed on coupons that exhibited low SIR. Figure 2.13 shows a scanning electron microscope view of a leakage path between two traces of a comb pattern exposed to 40X of flux. Copper corrosion was also observed on these combs. Thus, the SIR results supported the accelerated aging results, and demonstrated that heavy deposits of flux residue cause high leakage currents.

The effect of Flux A on the SIR is representative of other LSFs tested. Figure 2.14 shows the dependence of SIR on flux quantity for eight different LSFs. It

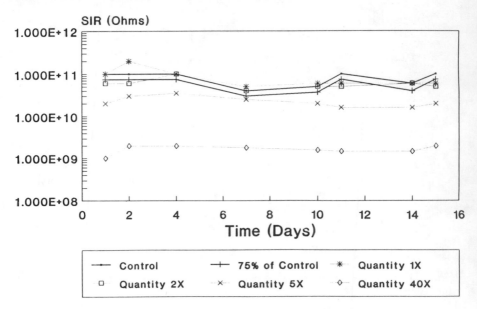

Fig. 2.12.   SIR data for comb patterns soldered with Flux A. The SIR is inversely proportional to the amount of flux applied (Guth 1989a).

is clear that, although all show the same trend, the SIR for some fluxes are more strongly dependent on flux quantity than others.

The SIR results for combs face up during soldering, as required in the Bellcore specification, showed that the flux residue on the topside of a board is more potentially corrosive than that on the bottom. The SIR results of this test indicate that top side flux residues can be more detrimental than those on the bottom side of a circuit board; also, visible corrosion was evident on combs with excessive flux residues.

The consequences of leakage currents can be significant, from signal termination to intermittent signals or signal alteration. The dependency of the SIR on flux quantity demonstrates a need for both process monitoring and process control. Because of the effect of excessive flux residue on the board surface, the flux deposition should be precisely controlled.

Drawbacks of Low Solids Fluxes.   The drawbacks of using low solids fluxes compared to higher solids content fluxes are summarized below. (They are discussed in greater detail in subsequent sections.)

- Quantity is critical to ensure minimal residue on the one hand, and sufficient fluxing activity on the other.

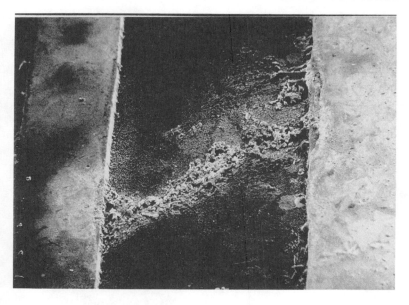

Fig. 2.13.   Scanning electron microscope view of the area between two traces on a comb pattern exposed to a 40X quantity of Flux A (Guth 1989a).

- Specific gravity is difficult to control.
- The effect of water absorption can be more pronounced than with fluxes of more typical solids content.
- Foaming is difficult.
- Process window may be narrower than when using fluxes with more typical solids content

Because of these drawbacks with LSF, the selection of application equipment and solder process parameters is important. If these are chosen wisely, the use of LSFs can yield enormous benefits.

2.3.5.   Controlled Atmosphere Soldering

Controlled atmosphere soldering has also been called inert gas or fluxless soldering. Inert gas soldering or fluxless soldering are misnomers, since more than inert gases are used, either reducing gases or vacuum atmospheres. Furthermore, liquids consisting of a flux solvent and activator are commonly used.

Recently, several companies have developed wave soldering systems that purportedly do not require the use of fluxes, and subsequently do not require postsolder cleaning. Another advantage which is cited is the decreased amount of solder dross.

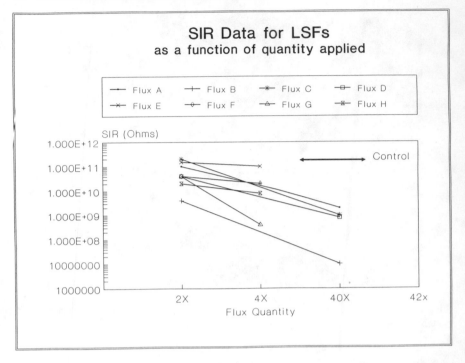

Fig. 2.14.   SIR data for eight low solids fluxes as a function of quantity applied (Guth 1989b).

One system, in particular, uses a combination of nitrogen and formic gases to create an oxygen-free environment. However, to insure proper through hole soldering, a liquid consisting of an organic acid and an alcohol is applied by an ultrasonic atomizer. Although this material is not called a flux, it is by all definitions a flux and should be treated as such in terms of its impact on reliability. Another controlled atmosphere soldering system relies on vacuum pumps and chambers to eliminate oxygen in the wave soldering module; this system is still in the development stage. Although these new processes sound promising, further evaluations of the reliability impact of the gases and liquids on the printed wiring assemblies need to be performed.

2.4.   APPLICATION METHODS

There are a number of ways to apply fluxes to printed circuit boards for wave soldering, the most common being wave, foaming, and spraying. These are discussed in the following sections, with an emphasis on the attributes necessary for low solids fluxes. The end of this section summarizes the applications issues for low solids fluxes.

Fluxes can also be applied by brushing, rolling, and dipping, but these are not as popular (Chung, Jancuk, and Corsaro 1983). Brushing is the simplest method, although it lacks uniformity and quantity control. Selective areas can be fluxed with brushing; however, the brushes can be difficult to clean and maintain. A roller can be used to coat a circuit board with no components, providing good uniformity, but the coating is less uniform when component leads are present. Dipping is simple but neither uniformity nor quantity can be controlled.

2.4.1. Wave

A wave fluxer is similar to the solder wave itself; the flux is pumped continuously through a wide nozzle located in an open reservoir or pool of flux. It is quite simple to operate and maintain. Another advantage is that virtually any liquid flux can be applied this way. Because a wave fluxer can apply excessive quantities of flux, it is sometimes followed by brushes or an air knife to remove excess.

For the wave fluxer, uniform and minimal application is difficult, and pushing flux up onto the top side of a circuit board is common. Maintaining the proper wave height and uniformity is necessary to avoid too much or too little flux on the printed circuit board. Pushing flux onto the top side of the board can be viewed as an advantage or a disadvantage; either it is good because the plated through holes (barrels) are filled with plenty of flux, promoting top side fillets, or it is bad if the top side flux residues can be deleterious and not easily removed with cleaning. Also, there are some open components that could be harmed if they came into contact with some fluxes. Consequently, selecting the appropriate fluxer for a particular flux is important.

With a waver flux, maintaining the original specific gravity (s.g.) requires adding solvent, since the flux is in an open reservoir, and exposed to the ambient. With most fluxes, the specific gravity (s.g.) can be controlled adequately with an automatic density controller that monitors the s.g. and adds additional solvent or flux when the s.g. gets out of range. However, with low solids fluxes, this issue of changing specific gravity can be detrimental, since small changes in the flux solvent content can greatly change the composition due to the small amount of solids (see section 2.4.4 of this chapter).

2.4.2. Foam

A foam fluxer consists of an open reservoir of flux which contains a submersed porous stone through which air is pumped. With the right combination of air pressure, nozzle size and shape, and surface tension of the flux, air bubbles are created in the pool of flux, and pushed through a nozzle, creating a foam of flux at the surface. Maintaining a level and uniform foam head with fine bubbles can sometimes be difficult since it depends on a number of factors. In recent years, modifications have been made to foam fluxers to make it easier to foam low solids fluxes; one manufacturer has designed a foamer with a duel aerator stone and an enlarged nozzle opening.

As with wave fluxers, a concern with foam fluxers is their tendency, when not adjusted properly, to bubble flux up onto the top side of a circuit board. An issue with LSFs is that because of the extremely low solids content, these fluxes can be quite difficult to foam. Foaming agents are added which could present a product reliability problem since they may be hygroscopic. The selection of a low solids flux for foaming applications should be made only after a careful evaluation of tests for corrosion and the effect of the flux residue on surface insulation resistance. Toubin (1989) suggests a number of elements to improve low solids flux foaming including the shape of the nozzle, the porosity of the stone, the air stream contents and pressure, and the geometry of the flux tank.

Like a wave fluxer, a foam fluxer causes changes in the specific gravity of the flux since the flux is not only exposed to the ambient but is aerated. So either a density controller is necessary or, in the case of LSFs, caution is suggested. Another especially important issue with foam fluxes is the possibility of water absorption by the isopropanol or other hygroscopic flux constituents. The addition of water can increase solder defects, cause spattering and solder balls, and can give a false confidence in s.g. maintenance. That is, the addition of water can lead one to believe that the s.g. is correct when, in fact, it is only in the right range due to the presence of an impurity (water).

Because of the nature of both wave and foam fluxers, fluxes applied with these methods can easily become contaminated with water, dirt from the printed circuit boards, conveyor fingers, or fixtures.

2.4.3. Spray

There are several different spray applications methods, high velocity spray, ultrasonic spray, and rotating drum spray, all with advantages and disadvantages, which are discussed in detail below.

High Velocity Spray   A high velocity spray fluxer, which can have a nozzle similar to one on a spray cleaner, can propel flux onto the top side of the circuit board. Moreover, units with large spray patterns can coat the inside of solder machines, and can create maintenance problems. Typical nozzles create spray patterns with relatively higher concentrations in the center, making uniform application difficult. Also, because of the nozzle design and the evaporating solvent, typical spray nozzles are easily clogged. Sometimes, a solvent after-spray is discharged after the flux spray to prevent or, at least, minimize clogging (Chung, Jancuk, and Corsaro 1983).

Ultrasonic Spray.   Because of the importance of controlling the deposition of low solids fluxes, a novel means of applying LSFs was developed and patented (Fisher, Guth, and Mahler 1989).

The theory behind ultrasonic atomization has been documented (Berger 1988). Vibrating at high frequency, a standing wave is created in the nozzle with sufficient energy to disintegrate the liquid flux into a fine mist. The frequency for the nozzle is dependent on the geometry of the orifice itself. And the orifice size is selected to deliver acceptable flux quantities to circuit boards. While the flow rate controls the amount of flux that will be dispensed, the power into the nozzle controls the amplitude of the sound waves into the nozzle, affecting the droplet size. Optimally, small droplets are created so that the most uniform application results. A low velocity air stream directs the spray mist upwards. A drawing of the flux mist being generated, and subsequently shaped in the mixing chamber and directed upwards is shown in Figure 2.15.

The ultrasonically controlled spray fluxer offers several advantages over other fluxers, especially pertinent for low solids fluxes. First, flux quantity can be controlled by adjusting the flux flow rate, ensuring product quality and reliability. Second, this is a closed system, preventing the evaporation of the flux solvent, and thereby assuring constant flux composition, especially important for low solids fluxes. Moreover, the closed system prevents water absorption. There is an adjusted spray interval timer on this fluxer so that waste is minimized. Because of the nature of the low velocity flux mist, a minimal amount of flux is deposited on the top side of the board, important for LSFs which can degrade the SIR more severely when not heated fully to soldering temperature. Minimizing the amount of flux that gets on the topside of the board must be balanced with the need to get enough flux in the plated through holes to ensure proper hole fill.

Details about the facility design and the process characterization for this type of application method have been documented (Guth 1989a).

Rotating Drum Spray.   Another spray method that has been in existence for a number of years has recently been touted as an application method for low solids fluxes. It consists of a partially immersed mesh drum rotating in a flux tank. Located on its center axis is an air knife, which, in combination with the rotating drum, creates a fine flux spray. The quantity of flux that is sprayed depends on the mesh size, the rotation speed, and the air velocity of the air knife. So, the quantity of flux can be controlled with this application method. However, the issues regarding control of specific gravity and possibility of water absorption exist with this method since there is an open flux tank.

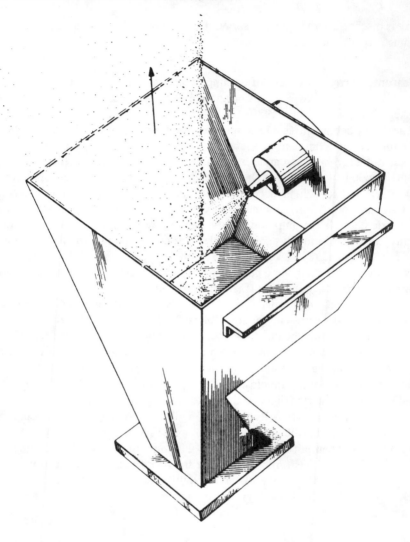

Fig. 2.15. Ultrasonically atomized flux being shaped and directed in mixing chamber (Fisher, Guth, and Mahler 1989).

2.4.4. Application Issues for LSFs

Each of the commercially available application techniques has its own list of advantages and disadvantages. From a flux quantity standpoint, a wave fluxer applies more than a foam fluxer which applies more than insert a spray fluxer.

Already discussed is the fact that the amount of LSF applied may affect the

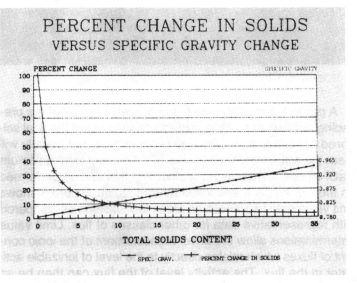

Fig. 2.16.   Percent change in solids versus specific gravity as a function of total solids content (Deram 1989).

SIR and ultimately the reliability of a printed circuit board. Consequently, it is important to control the quantity of flux applied. Normally, a board's cleanliness is controlled by the efficiency of the cleaning material and cleaning process; here, the control must be directed at the flux application process step since there is no postsolder cleaning. As important is the amount of unwanted flux that is deposited onto the top side of the printed circuit board since partially heated low solids flux residues may be corrosive.

Another aspect that is extremely important in low solids flux application is the control of the composition. Since these fluxes have so little solids, their specific gravity is quite similar to that of isopropanol. In fact, for most automatic density controllers, the s.g. of an LSF is indistinguishable from that of isopropanol. Consequently, an application method that stores the flux in a closed container is preferred over those with open tanks, avoiding potential problems with changes in s.g. The fact is that small changes in solids content greatly influences the performance of low solids fluxes. It has been shown that the percent change in solids increases dramatically with decreasing solids content, even though the change in specific gravity is linear (Deram 1989). This relationship is illustrated in Figure 2.16. Deram also reported that low solids fluxes have exceeded twice their original concentration when controlled only by specific gravity. A titration for acid number is a preferred control for low solids fluxes; this type of controller is commercially available.

Regarding the issue of water absorption, this effect can be magnified with a low solids flux, since a small amount of absorbed water can be much higher in proportion to the solids in an LSF than the solids in a traditional flux. Klein Wassink (1984) illustrates the effect of 5 wt.% of water on the specific gravity of a traditional rosin flux, increasing it by approximately 0.014. This absolute magnitude of increase has a more significant effect when using LSFs.

2.5. MONITORING TECHNIQUES

To ensure that the flux material remains consistent over time and that appropriate amounts of flux are being applied to the printed circuit boards, several monitoring techniques can be used. Techniques discussed below are specific gravity, acid number (titration), and weight measurements, followed by a discussion of LSF monitoring.

2.5.1. Specific Gravity

Monitoring specific gravity (s.g.) is the traditional method used for maintaining the proper solids concentration. By definition, specific gravity is simply the ratio of the density of a substance to that of water. It can, therefore, be used to determine the solids content, assuming a closed system. That is, ideally, nothing else should alter the specific gravity; however, in reality, water absorption due to the hygroscopic nature of some of the flux components (including isopropanol) will obviously affect the s.g. reading, giving a false indication of solids content. It is quite simple to monitor the s.g. either manually or automatically for most fluxes. However, as it has already been discussed in Section 2.4.4, this method is not precise enough for low solids fluxes.

2.5.2. Acid Number (Titration)

The acid number for a flux involves an acid-base titration. The addition of a certain amount of acid (flux) to a basic solution causes the pH indicator that is present to change colors. That is, the acid number is a value determined by the amount of acid necessary to cause a color change. It is a more precise measure of the flux components than specific gravity determination. Another advantage of acid number is that it is unaffected by the water content of the flux, unlike specific gravity. Test kits are available from the flux manufacturer so this test can be done on the production floor (Deram 1989).

2.5.3. Weight Measurements

Weighing the amount of flux solids that are applied to a printed circuit board involves weighing the board before and after flux application, allowing time for the solvent to evaporate so that a steady weight can be achieved. This technique monitors the amount of flux that is being applied to the board, a quantity that is affected by both the composition of the flux and the application method. This method was used in the studies of low solids fluxes discussed in Section 3.4.3. The amount of flux in terms of nonvolatiles per unit area of substrate can be calculated.

2.5.4. Monitoring for LSFs

The specific gravity (s.g.) of these fluxes is extremely difficult to control with an automatic density controller since the s.g. is almost the same as the solvent or thinner itself. Acid number (titration) is a more precise measure of the flux composition. However, while titration will monitor the flux composition, it will not monitor the amount of flux that is being applied which is important for low solids fluxes. Although it can be cumbersome to do, weight measurements can give an indication of the quantity of flux applied.

Because it is difficult to monitor low solids flux composition, it is suggested that an application method with flux in a closed reservoir be used so that the solvent cannot evaporate and water cannot be absorbed.

2.6. PROCESS ISSUES

There are a number of process issues associated with the type of flux used. These issues include flux residue, solder ball formation, top side fillet formation, and conformal coating compatibility, all of which are discussed below.

2.6.1. Flux Residue

Depending on the flux type, flux residue may or may not be a reliability concern. The more aggressive fluxes leave behind more corrosive residue and should therefore be removed properly. Less active fluxes or those that meet certain tests requirements without removing their postsolder residue may be removed for aesthetic reasons, or to avoid problems with testing, solder balls, or conformal coating.

2.6.2. Solder Ball Formation

Material incompatibilities can cause excessive solder balls and/or other solder type residue. The number of solder balls is a function of the type of flux used and the type of solder mask on the printed circuit board. The presence of solder balls becomes a real issue when there is no postsolder cleaning (e.g., some rosin fluxes, low solids fluxes, controlled atmosphere soldering) which could remove the balls along with the other residue. In comparisons of a number of solder mask types and flux types (rosin, SA, WSF, and LSF), some industry experience indicates that the number of solder balls is greater with dry film solder masks and with low solids fluxes. Work has been done showing that the following factors can affect solder ball formation: dry film solder mask type, printed wiring board cleanliness, surface roughness, and flux type (Khutoretsky 1987). Khutoretsky's study included dry film solder masks and both water soluble and rosin fluxes; the results indicated that both the flux residue and the solder mask polymer can retain solder balls on the board surface. Photomicrographs in Khutoretsky's paper illustrate this "connecting chain" phenomenon.

2.6.3.   Top Side Fillet Formation

Top side fillets are required in some specifications, and the definition of which varies amongst specifications. Some specifications simply require positive wetting on both the component leads and the barrel, and a certain amount of barrel ill, whereas other specifications might require a fillet on the top side that looks almost identical to that on the bottom side. The exact amount of solder necessary to create a reliable joint is debatable and obviously is affected by the conditions to which a printed circuit board will ultimately be exposed. That is, an electronic product which is made for military aircraft will be exposed to much harsher temperature and vibration conditions than a household consumer product.

The formation of sufficient fillets on the component side of the printed circuit board is affected by the board and component temperatures through the solder wave, their solderability, board thickness, the component-to-hole aspect ratio, the solder wave height, the conveyor speed, the flux application method, the activity of the flux, and the solids content of the flux. If the circuit board and components are not adequately heated, the solder will not rise properly in the holes. Moreover, if solderability is poor or the activity of the flux is not adequate, the solder will not readily wet. Differences in top side fillets have been observed between a 37% solids rosin flux and a 15% solids rosin flux; additionally, other parameters also affected top side fillet formation in this study including conveyor speed, height of flux foam, and the preheat profile (Wittmer 1989). For LSFs, problems with top side fillet formation can be pronounced.

2.6.4.   Conformal Coating Compatibility

Conformal coating is commonly applied after soldering to printed circuit boards which will be exposed to harsh or extreme conditions. These include military applications and automobile electronics. Some flux residues underneath these coatings can cause blistering, peeling, or measling. Any moisture or hygroscopic residues underneath the coating can cause these problems. Obviously then, coating compatibility with a flux and its residue should be taken into account. Although rosin fluxes are usually associated with product requiring conformal coating, reports by members of the electronics industry indicate that coatings exist that are compatible with other flux types including WSF and LSF; cleaning before the coating process is usually necessary and depends on the type of flux used.

2.7. NONLIQUID FLUXES

2.7.1. Core Solder Material

Instead of liquid fluxes that are used for wave soldering, hand soldering is usually performed with solder wire that contains a core or several cores of flux material. This material is solid or semi-solid at room temperature, but flows and is active at soldering temperatures to promote adequate wetting of the solder. Rosin-based core solder has been the standard in the electronics industry but in the last few years there has been an increase in water soluble core flux and, more recently, low solids flux core solders. Activators may be organic acids and/or organo-halides.

Core materials can be tested just like liquid fluxes by obtaining the core material from the manufacturer or extracting it from the solder wire. However, although some specifications state that the core material may be removed by melting the solder wire and then testing, this practice is emphatically not recommended. When the solder is melted, the core material is also heated, and can thus react with the solder or simply undergo changes by itself. As a consequence, the material tested will have undergone one more heating step than allowed or required in tests such as copper mirror or surface insulation resistance.

2.7.2. Solder Paste Material

Solder pastes are often used for attaching surface mount devices. The paste is stenciled on pads on the printed circuit board, the components are placed onto the pads, and the whole assembly is heated so that the solder paste is reflowed. The solder paste is a combination of solder powder, rheology modifiers, and normal flux constituents such as a high temperature solvent, a vehicle, and activators. The difficulty in finding a good solder paste is due to the number of requirements imposed upon these materials. That is, along with acceptable flux and solder materials, a paste must also be formulated to include other properties such as tackiness, printability, and minimal slumping. Therefore, solder pastes are usually tested like other fluxes for chemical reliability purposes, and then undergo further tests to ensure the presence of certain physical properties. The electronics industry specification for solder pastes is IPC-SP-819.

Solder pastes based on any of the four major flux types exist. However, most solder paste used today is rosin based with minimal halide activators because of cleaning concerns and questions regarding the need for higher activity. Because of the additional materials in solder pastes and the localized concentrations of paste residue compared to wave soldering flux residue, cleaning is more difficult and more of a concern than cleaning after wave soldering.

Removal of solder paste residue is sometimes, but not always, necessary for the same reasons as for liquid fluxes. However, with the increasing problems with cleaning, lower solids content solder pastes are being developed. Water soluble solder pastes are available for applications such as mixed technology where water soluble liquid flux and solder paste residue will be removed in a water cleaning system. The concern is that the residue be removable after the heating of soldering and the electrical properties (SIR) of the assembly have not been excessively degraded. Even though the flux-type ingredients may be water soluble, the rheological agents are not necessarily so. Hwang (1989) offers an extensive discussion about solder pastes in general.

2.7.3. Specifics for LSFs

For liquid fluxes, core solders or solder pastes, less solids means less residue. An effective way to reduce flux residue with cored wire solder is to use smaller core sizes. Typically the standard 3% flux (by weight) can be reduced to 2%, 1%, or 0.5%. With less flux being used, solderability and heat control become more important to insure that reliable solder connections are made. For solder pastes, it is difficult to truly make a lower solids content paste, and also maintain the proper physical properties. That is, the less nonmetal material that exists, the more solder powder that is present; obviously, the printability and tackiness properties will then be affected. Thus, it seems a paste that leaves less or no residue should contain the usual proportion of solder powder (approximately 90 wt.%), but include other materials that decompose during reflow. Low solids pastes reflowed in controlled atmospheres are being investigated (Morris 1990).

2.8 IMPORTANCE OF SOLDERING PARAMETERS

Regardless of the soldering method, there are certain parameters that can affect the ability to remove the post solder flux residue, specifically those that affect both the maximum temperature and the period of time at high temperature to which a printed circuit board is exposed. Usually, the longer a board is exposed to a high temperature, the more difficult it is to remove the flux residue. Furthermore, for all types of fluxes, the soldering efficiency can be affected by the soldering parameters.

2.8.1.   Wave Soldering

Many parameters control the amount of time at high temperature and the maximum temperature experienced by printed circuit boards during wave soldering.
They include the preheat temperature, the length of the preheat stage, the solder
pot temperature, the conveyor speed, the size and/or number of solder waves,
the conveyor speed, the height of the solder wave, and the shape or type of
solder wave. As important as the ability to remove the postsolder flux residue
are other issues which are affected by the soldering parameters such as minimizing
the number of solder defects as well as the thermal shock to components, especially surface mount ceramic capacitors. Many articles have been written about
troubleshooting wave soldering problems, including a recent one by Elliott (1989).

2.8.2.   Hand Soldering

To make reliable solder joints consistently with a hand soldering iron is not a
simple task. Important parameters that can affect the soldering and the remaining
core solder flux residue include the shape and size of the iron, the temperature,
the iron's tip material, the power consumption, the amount of time the iron
contacts the joint, and the general technique of the operator. Because this is a
manual operation, hand soldering can sometime be the most difficult to control
and to maintain consistency. With surface mount component soldering, technique
is even more critical. Special tips and other tools have been designed to handle
the specific problems with SMT (Johnson 1989).

2.8.3.   Reflow Soldering

Most electronic manufacturers use infrared (IR) heating to reflow solder paste
when joining surface mount components. Important parameters to optimize include the temperature at each stage in an inline machine, the uniformity of the
temperature across the width of the conveyor, and the conveyor speed. As with
the other heating processes, components with large thermal mass can greatly
alter the temperature of the entire assembly and, as a result, cause soldering
problems. Elliott (1989) offers advice in improving an IR reflow process.

2.8.4.   Specifics for LSFs

For wave soldering, it was at one time believed that a lower preheat or none at
all would be beneficial for low solids fluxes since the solids content was so low.
For most LSFs, this has been proven false; higher preheats are necessary. Wade
(1989) suggests increasing the preheat 20 to 30 degrees. It is important to set
the conveyor speed such that it is not so slow that all the flux activity is lost

before exiting the wave and not so fast that the flux is not adequately displaced by the solder wave, leaving excessive flux residue. In general, electronics manufacturers have found a smaller process window exists for low solids fluxes, so there is less room for error.

Hand soldering with low flux percentage core solders has been found to be more difficult than with traditional rosin core solders. Because there is much less flux in comparison, there is a tendency to lose all the flux before the joint is made. That is, operators should be retrained and shown that minimizing the heating of the joints is beneficial. The core solder wire vendors sometimes also recommend a lower soldering temperature. Another issue for hand soldering is tinning the tips, since the low flux percentage is insufficient for tinning, unlike the traditional rosin core solder wire. So the right combination of low flux percentage and operator efficiency has to be established for reliable soldering.

There is not much experience with reflow of low solids solder paste. However, it has been seen with the few commercially available low solids solder pastes that a fast rise in temperature and fast cool-down are preferred over a slower heating and cooling. Thus, a reflow process such as condensation reflow (also called vapor phase reflow) may be better than IR reflow. If the heating and cooling are too slow, the low solids solder paste dewets from the surface to be soldered because all the activator is used up prematurely. Another solution to this problem is the use of controlled atmospheres during reflow (Morris 1990).

2.9. SUMMARY AND TRENDS

2.9.1 Summary

This chapter has discussed the types of requirements and tests that are used to qualify fluxes as well as the types of fluxes which are currently available along with their advantages and disadvantages. In addition, flux application methods, flux monitoring techniques, process issues, as well as issues regarding nonliquid fluxes were covered. From the information in this chapter, it should be clear that selecting an appropriate flux is not an easy task nor should a cleaning material be chosen without first scrutinizing the type of flux that is necessary.

2.9.2. Trends

It is probably obvious to everyone by now that the major push is away from chlorofluorocarbon (CFC) cleaning, and if possible toward the elimination of cleaning altogether. There are several possible ways to avoid cleaning including the use of noncorrosive rosin fluxes, the use of low solids fluxes, and the use of controlled atmosphere soldering. According to Szymanowski (1989), controlled atmosphere (inert gas) soldering and low solids fluxes offer the most attractive alternatives to CFCs. In those cases were cleaning is still necessary, HCFCs, terpenes, or aqueous media are the choices. However, HCFCs are an interim solution only, since their use may be restricted in the future. Presently, there are still no soldering methods that are truly "fluxless" and there may never be any that do not use some kind of fluxing material. But with low solids fluxes and controlled atmosphere soldering, at least, we are getting closer to that dream.

In the future, improvements in wave soldering low solids fluxes will be seen as well as improvements in low solids solder core wire and low solids solder pastes.

Other types of fluxes can and will still be used with appropriate non-CFC cleaning media. A preferred cleaning medium for rosin flux residue removal is semi-aqueous and terpene-based (see Chapter 5 of this book.) If a completely aqueous system is preferred, water soluble flux can be used if the assembly allows complete removal of the flux residue with water. However, the addition of a cleaning process brings additional expense, maintenance, capital equipment, and floor space. With the every increasing restrictions on the wastes which can be expelled from factories, minimizing or eliminating cleaning altogether is certainly the preferred direction and the trend.

REFERENCES

Aspandiar, R., Piyarali, A., and Prasad, P., 1986. "Is OA OK?" *Circuits Manufacturing,* April, 29–36.

Bellcore. 1987. TA-TSY-000078, Issue 2. "Proposed Revision of TR-TSY-000078: Generic Physical Design Requirements for Telecommunications Products and Equipment." Red Bank, NJ: Bell Communications Research, Inc.

Bellcore. 1988. TR-TSY-000078, Issue 2. "Generic Physical Design Requirements for Telecommunications Products and Equipment." Red Bank, NJ: Bell Communications Research, Inc.

Berger, H. 1988. "Ultrasonic Nozzles Atomize Without Air." *Machine Design,* July 21, 58–62.

Chan, A. S. L., and Shankhoff, T. A. 1989. "Interrelating Surface Insulation Resistance Test Patterns." *Circuit World,* 15(4):34–38.

Chung, B. C., Cassidy, M. P., and Graham, G. W. 1983. "Evaluation of Flux Performance, Cleaning and Reliability." *The Western Electric Engineer* (1):30–39.

Chung, B. C., Jancuk, W. A., and Corsaro, V. A. 1983. "Aqueous Detergent for Removing Rosin Fluxes." *The Western Electric Engineer* (1):62–68.

Danford, A., and Gallagher, P. 1987. "SMD Cleanliness in an Aqueous Cleaning Process." *Nepcon East 1987,* 245–255.

Deram, B. 1989. "Considerations for Use of No-Clean Fluxes in Soldering PCB's. *Electronic Manufacturing,* February, 32–34.

Dickinson, D. A., Guth, L. A., and Wenger, G. M. 1989. "Advances in Cleaning of Electronic Assemblies." *Nepcon West 1989.*

Elliott, D. 1989. "How to Avoid Problems with Wave Soldering and IR Reflow." *Surface Mount Technology,* October, 47–54.

Fisher, J. R., Guth, L. A., and Mahler, J. A. 1989. "Method and Apparatus for Applying Flux to a Substrate." U.S. Patent 4,821,948, April 18, 1989, assigned to AT&T.

Gorondy, E. J. 1984. "Surface Insulation Resistance—Part I: The Development of an Automated SIR Measurement Technique." *IPC Technical Paper* IPC-TP-518.

Gorondy, E. J. 1988. "Surface/Moisture Insulation Resistance (SIR/MIR)—Part III: Analysis of the Effects of the Test Parameters and Environmental Conditions on Test Results." *IPC*-TP-825 *Technical Paper*.

Guth, L. A. 1988. "Low Solids Flux Material Characterization Studies." *IPC Technical Paper* IPC-TP-735.

Guth, L. A. 1989a. "Low Solids Flux Technology for Solder Assembly of Circuit Packs." *Proceedings of the Electronic Components Conference*, 748–753.

Guth, L. A. 1989b. "Flux Selection and Application to Negate Cleaning Needs." Presented at Nepcon East, Boston, MA, June 14.

Hayes, M., and Hood, C. 1989. "Alternative Cleaning Methods: One Possibility." Interview in *Printed Circuit Assembly*, September, 33–35.

Hwang, J. S. 1989. *Solder Paste in Electronics Packaging*. New York: Van Nostrand Reinhold.

IPC. 1988. ANSI/IPC-SF-818, "General Requirements for Electronic Soldering Fluxes." Lincolnwood, IL: Institute for Interconnecting and Packaging Electronic Circuits (IPC).

Johnson, R. O. 1989. "Solderability Issues for Hand Soldering." *Printed Circuit Assembly*, June, 30–32.

Kenyon, W. G. 1986. "Synthetic Activated (SA) Flux Technology: Development, Commercialization, Benefits and Future Applications." *Proceedings of Internepcon Japan '86*.

Khutoretsky, M. 1987. "An Evaluation of Solder Balls on Printed Wiring Boards with Dry Film Solder Resist." *Proceedings of ASM's Third Conference on Electronic Packaging: Materials and Processes & Corrosion in Microelectronics*, 225–233.

Klein Wassink, R. J. 1984. *Soldering in Electronics*. Ayr, Scotland: Electrochemical Publications, Ltd.

Morris, J. R. 1990. "No clean Alternatives for Solder Paste Reflow." *Proceedings of SMART VI Conference*, (2):97–106.

Reagor, B. T., and Russell, C. A. 1986. "A Survey of Problems in Telecommunication Equipment Resulting from Chemical Contamination." *IEEE Transactions on Components, Hybrids, and Manufacturing Technology*, **CHMT-9**(2)209–214.

Rubin, W. 1982. "Some Recent Advances in Flux Technology." *Brazing and Soldering*, Spring (2):24–28.

Sholley, C. 1989. "New Fluxes Eliminate Cleaning of PCB's." *Electronic Manufacturing*, March, 32–33.

Szymanowski, R. A. 1989. "Fluxing Options for CFC Elimination." *IPC Technical Review*, April–May, 19–23.

Toubin, A. 1989. "Low Solids Content Fluxes." *Circuit World*, **15**(2):17–18.

Wade, R. L. 1989. "A View of Low-Solids Fluxes." *Printed Circuit Assembly*, March, 31–34.

Wargotz, W. B., Guth, L. A., and Stroud, C. V. 1987. "Quantification of Cleanliness Beneath Surface Mounted Discretes Assembled by Wave Soldering." *Printed Circuit World Convention IV*, WCIV-72.

Wittmer, P. 1989. Conversation with the author. Unpublished Magnavox data.

Zado, F. M. 1983a. "Effects of Non-Ionic Water Soluble Flux Residues." *The Western Electric Engineer* (1):40–48.

Zado, F. M. 1983b. "Increasing the Soldering Efficiency of Noncorrosive Rosin Fluxes." *The Western Electric Engineer* (1):22–29.

Zado, F. M. 1984. "Electrical/Electronic Reliability Considerations in Modern PWB Manufacturing and Assembly Operations." *Printed Circuit World Convention III* WCIII-70.

3
Solvent Defluxing of Printed Wiring Board Assemblies and Surface Mount Assemblies: Materials, Processes, and Equipment

J.K. "Kirk" Bonner

3.1. THE NEED FOR CLEANING

In this chapter the chief topic of discussion will be postsolder cleaning of printed wiring board assemblies and surface mount assemblies. Given the title of this chapter, the focus will be even narrower, and the topic will be restricted to solvent cleaning only. Cleaning following the solder operation is also known as defluxing. However, cleaning in relation to printed wiring board assemblies and surface mount assemblies has a broader meaning. A cleaning operation can potentially follow any process operation. If one is referring specifically to that cleaning operation following the soldering process, the term *defluxing* will be used.

The term *cleaning* will be used to refer to the removal or extraction of contaminants from the surface of the printed wiring board, printed wiring board assembly, or surface mount assembly. Solvent cleaning refers to the use of some organic or carbon-containing material in conjunction with a piece of equipment for effecting the removal of contaminants. Again, after the soldering operation, the equipment used is referred to as defluxing equipment. To put the whole issue of cleaning in perspective, it is necessary to review the manufacturing operations involved in bare board, printed wiring assembly and surface mount assembly production.

3.1.1. The Bare Board and Assembly Process—An Overview

The printed wiring board (PWB), otherwise known as a printed circuit board, still continues to be the major interconnection substrate device for electronic components. However, because the technology of substrates is evolving at an accelerated pace along with component technology, it has been proposed that the substrate be designated the packaging and interconnect (P/I) structure. Whether this term will become widely accepted is still a moot point.

The fabrication of the PWF or P/I structure means, in the trade, the sum total of manufacturing steps required to produce a substrate. This substrate can have circuit traces on its surface, buried layers (power, ground, signal) if the PWB is a multilayer board, through holes for conventional throughhole components (THCs), vias, etc. The substrate may have solder mask applied over the circuit traces. The laminate materials used as the backbone of the PWB can be standard G-10/FR-4 glass epoxy, paper-filled with glass sandwich epoxy, modified epoxy, polyimide, Teflon, or some other material. But the fabrication (or fab) process does not include component mounting or insertion and subsequent attachment of these components to the PWB. This latter process is known as assembly. The PWB with its components attached to it by a metallurgical process is referred to as a printed wiring assembly (PWA), a PWB assembly, a printed circuit board or module.

The fabrication process itself is complex. Typically, using an exposure machine the circuit artwork is imaged onto the copper-clad laminate to which photoresist has been attached. The excess photoresist is then removed. Unwanted copper is etched away and the remaining photoresist removed to reveal the traces desired for the final board. If this is to be an innerlayer of a multilayer board, the copper will not be plated. If it is to be an outerlayer, it may be solder plated. However, solder mask over bare copper is becoming much more common. The major processes are: imaging, drilling, etching, plating, and routing. But it should be emphasized that there are numerous lesser operations that take place also such as deburring, developing, stripping, rinsing, etc. In addition, the plating operations and etching operations are not simple one-step operations but involve a multiplicity of steps, many of them quite complex as regards the chemistry. Inspection processes also play a critical role in producing the final PWB. List 3.1 gives a typical series of manufacturing operations for producing a solder-plated PWB.

The assembly process involves mounting or inserting the components on the bare PWB and attaching them permanently to the PWB, generally by the use of a soldering process. Once the components are permanently attached to the PWB, the piece is now referred to as a printed wiring assembly (PWA). If surface mount components (SMCs) are employed during the assembly process, the resulting assembly is commonly called a surface mount assembly (SMA). There are several types of SMAs. Type I has only surface mount components which are attached typically by a reflow method. The components may only be top side or both top and bottom side. Type II has both SMCs and THCs. Generally, the SMCs on the top side are attached using a reflow method and the THCs are attached using a wave soldering method. The THCs are normally only on the top side of the board, the SMCs may be top side only or top side and bottom

List 3.1. Typical Series of Manufacturing Operations for Producing a Solder Plated Double-Sided PWB.

1. Image copper-clad laminate
2. Drill through holes
3. Electroless copper plate PWB
4. Develop photoresist
5. Copper plate PWB
6. Tin/lead plate PWB
7. Strip PWB
8. Etch PWB
9. Size PWB to final dimensions (route, shear, or punch)
10. Mask off finger areas
11. Nickel/gold plate
12. Bevel finger contact area
13. Apply fusing fluid
14. Reflow solder
15. Electrical test
16. Apply solder mask
17. Final inspection
18. Package for shipping to customer

side. If they are on the bottom side, they are often attached using an adhesive followed by wave soldering the board. Type III again has both THCs and SMCs but the SMCs are mounted on the bottom side using an adhesive and then the entire board is wave soldered.[1] List 3.2 shows a typical series of manufacturing operations for attaching THCs to PWBs. List 3.3 shows a typical series of operations for producing a Type I SMA; List 3.4 shows a typical series of operations for producing a Type II SMA; and List 3.5 a typical series of operations of producing a Type III SMA. A final step, application of a conformal coating, is optional in all cases.

List 3.2. Typical Series of Manufacturing Operations for Producing a PWB Assembly.

1. Insert throughhole components
2. Wave solder
3. Clean
4. Test for cleanliness
5. Test/repair

List 3.3. Typical Series of Manufacturing Operations for Producing a Type I Surface Mount Assembly.

 1. Screen print solder paste
 2. Place surface mount components
 3. Cure solder paste
 4. Reflow solder
 5. Clean (optional)
 6. Turn over to side 2
 7. Screen print solder paste
 8. Place surface mount components
 9. Cure solder paste
10. Reflow solder
11. Clean
12. Test for cleanliness
13. Test/repair

3.1.2.   Types of Contaminants

It should be obvious from the discussion above that PWBs during manufacture are exposed to a wide variety of potential contaminants. Among these are: photoresist residues, etching salts, plating salts, metallic slivers, drill debris, and if the PWB has solder plate that is reflowed, fusing fluids.

The PWB manufacturer may or may not clean his finished PWBs prior to shipping. If he does not, this may leave a considerable amount of contamination on the bare board. In addition, human handling and airborne particulates can add to the burden of contamination found on the PWB.

When the PWB goes through the assembly process, it is exposed to another set of contaminants, arising chiefly from the fluxing/soldering operation. The fluxing operation produces a great deal of contamination on the final assembly, and this is the chief reason why assemblers deflux their assemblies. Since fluxing/soldering introduces such a wide variety of contaminants that are potentially harmful if left on the PWB assembly surface, it is relevant to discuss in some detail the chemistry of fluxes and solder pastes.

Fluxes.   The purpose of a flux is to strip the oxide barrier off a metallic surface and also any other barrier, such as that of a sulfide, thus exposing the virgin metal underneath. This stripping of oxides and sulfides off the surface makes the metal solderable. If the oxide barrier is not removed properly, it will inhibit the wetting action of the molten solder during the soldering process. Fluxing also insures a reduction of the surface tension between the solder and the surfaces

List 3.4. Typical Series of Manufacturing Operations for Producing a Type II Surface Mount Assembly.

1. Screen print solder paste
2. Place surface mount components
3. Cure solder paste
4. Reflow solder
5. Clean (optional)
6. Insert throughhole components
7. Turn over to side 2
8. Apply adhesive to side 2
9. Place surface mount components
10. Cure adhesive
11. Invert board
12. Wave solder
13. Clean
14. Test for cleanliness
15. Test/repair

to be soldered. This improves the wettability of the surface, facilitating the spread of the solder and again promoting good solderability.[2] In fact, the word *flux* is derived from the Latin *fluere*, to flow.

Flux manufacturers today produce a wide variety of fluxes. These can be categorized into fluxes based on rosin, synthetic activated (SA) fluxes, and water soluble (WS) fluxes. The water soluble (WS) fluxes are also sometimes referred to as organic acid (OA) fluxes although this is not advisable since many fluxes

List 3.5. Typical Series of Manufacturing Operations for Producing a Type III Surface Mount Assembly.

1. Insert throughhole components
2. Turn over to side 2
3. Apply adhesive to side 2
4. Place surface mount components
5. Cure adhesive
6. Invert board
7. Wave solder
8. Clean
9. Test for cleanliness
10. Test/repair

in the other categories also contain organic acids. Rosin fluxes themselves are no longer simple. Today flux manufacturers modify the organic acids (rosin acids: abietic, pimaric, etc.) in naturally occurring rosin to produce modified rosin fluxes to enhance the properties of the flux, e.g., to make it more resistant to heat to prevent thermal degradation and rosin polymerization.

In general, a flux consists of a solvent carrier such as isopropyl alcohol (IPA) which is generally quite volatile, a vehicle such as some organic material having a fairly high boiling point, and an activator system. In the case of rosin-based fluxes, the vehicle is the rosin. Activator systems based on amine hydrohalides are still quite common in fluxes used in the United States. An example of this is dimethylamine hydrochloride, $(CH_3)_2NH \cdot HCl$. Nonhalide fluxes typically contain dicarboxylic acids such as succinic or adipic acid as activators.

For rosin-based fluxes, which are used by almost all military contractors and also by some commercial manufacturers, the method of categorization has been based on MIL-F-14256. This military specification classifies rosin-based fluxes as rosin nonactivated (R), rosin mildly activated (RMA), rosin activated yet still meeting the Mil-spec requirements (RA-MIL), or if the flux is more active yet, as rosin activated (RA) or rosin superactivated (RSA). The Mil-spec requirement is based on several tests: (1) the resistivity of an aqueous extract of the flux, (2) the silver chromate paper test, and (3) the copper mirror test. An RMA flux is one containing a small amount of activator such that five water extracts of this flux have a mean specific resistivity of 100,000 ohm-centimeters or greater. An RMA flux will also not affect silver chromate paper, and it passes the copper mirror test. An RA-MIL flux is one which contains more activator. Typically, the mean specific resistivity of such a flux is between 100,000 ohm-centimeters and 50,000 ohm-centimeters. If the resistivity is less than 50,000 ohm-centimeters, the flux is considered highly active. Obviously, the more activator in the flux, the lower the specific resistivity of its aqueous extract.

It should be mentioned at this point that an alternate classification scheme for fluxes is recommended by the Institute of Interconnecting and Packaging Electronic Circuits (IPC). Essentially, many in the PWB industry believe that the classification scheme based on MIL-F-14256 no longer adequately reflects the present realities in flux formulation and flux development. The IPC is suggesting that flux types be categorized as either L (low), M (medium), or H (high). See Table 3.1. This recommendation is embodied in IPC-SF-818. The military is still using the classification of fluxes as distinguished by MIL-F-14256. Although this distinction highlights some of the differences between military PWB assemblers and commercial assemblers, there is a trend for more cooperation so that one set of standards may emerge for the entire industry.

The Limit Value of a Flux. A useful technique for characterizing the amount of activator in the flux is determination of the limit value of the flux.[3] This can

Table 3.1 Recommended Classification of fluxes (IPC-SF-818).

TYPE (FLUX AND FLUX RESIDUE ACTIVITY LEVEL)	TESTING REQUIREMENTS	
	TEST	REQUIREMENT
L	Copper mirror	No evidence of copper removal or partial removal of the copper mirror indicated by white background showing through in part of the fixed area.*
	Silver chromate or halide	Pass silver chromate test** or ≤0.5 percent halides.
	Corrosion	There shall be no evidence of corrosion. If a blue/green line occurs at the flux metal interface the area of corrosion shall be tested per 4.5.5.2 (silver chromate test) and must pass those test requirements.
M	Copper mirror	Partial or complete removal of the copper mirror in the entire area where the flux is located.
	Halide	≤2.0 percent halides.
	Corrosion	Minor corrosion is acceptable provided the flux or flux residue can pass the copper mirror or halide requirements for this category. Evidence of major corrosion places the sample or flux into the H category.
H	Copper mirror	Complete removal of the copper mirror in the entire area where flux is located.
	Halide	>2.0 percent halides.
	Corrosion	Evidence of major corrosion.

*Discoloration of the copper or reduction in the copper film thickness shall not be sufficient to remove the flux from this category.
**Failure to pass the silver chromate test requires performing and meeting the requirements of the halide test in order to be considered for category L.

be done using a simple production ionic contamination tester. See Figures 3.1 and 3.2. As was mentioned above, fluxes are characterized into three broad categories: rosin-based fluxes, synthetic activated (SA) fluxes, and organic acid (OA) or water soluble (WS) fluxes. Of these, the rosin-based fluxes are further broken down into rosin nonactivated (R), rosin mildly activated (RMA), rosin activated within military specification requirements (RA-MIL), rosin activated (RA), and rosin superactivated (RSA).

Fig. 3.1. Standard ionic contamination tester. (Courtesy of Alpha Metals Inc.)

Unfortunately, this type of qualitative categorization fails to describe the level of activator in a flux, which is a serious omission because the activator is responsible for good solderability. In addition, unremoved activator can lead to corrosion on the board.

Of course, a flux can be characterized using the procedure outlined in MIL-F-14256. However, the results will be in resistivity units: ohm-centimeters.

By using the fact that the amount of an activator can be determined by measuring its own content and by applying the mathematical concept of a limit value of a flux, one can quickly characterize the level of activator in a flux.

The limit value (LV) of a flux will equal the amount of ionic material, C_{Nacl}, expressed in micrograms of sodium chloride, divided by the flux volume, V_{Flux}, which is expressed in microliters:

$$\text{Limit Value } (LV) \text{ of flux} = \lim_{V_{\text{Flux}} \infty} (C_{\text{Nacl}}/V_{\text{flux}}).$$

Method of Determining the LV of a Flux. First calibrate the ionic contamination tester. This can be done as follows. By using a solution of sodium chloride, a

Fig. 3.2.   Ionic contamination tester with spray headers in test tank for SMAs. (Courtesy of Alpha Metals Inc.)

regression line can be generated to relate specific conductance, k, of the standard solution to the amount of sodium chloride, which should be measured in micrograms. If resistivity (ρ) values are used, convert the measurements to conductance by $k = 1/\rho$.

Using the technique of injecting different quantities of sodium chloride into the unit, a regression line will be obtained. The amount of sodium chloride is related to the conductance k linearly. That is, the plot should produce a straight-line curve. If the amount of sodium chloride is plotted against resistivity, a hyperbola will result. Because of this, conductance is a better measure.

$$k' = mC_{\text{NaCl}} + b.$$

In this equation

$$k' = k \times 10^2$$

and m and b are the regression constants. Of course, m is the slope of the regression line. Its value for a standard sodium chloride solution used with the

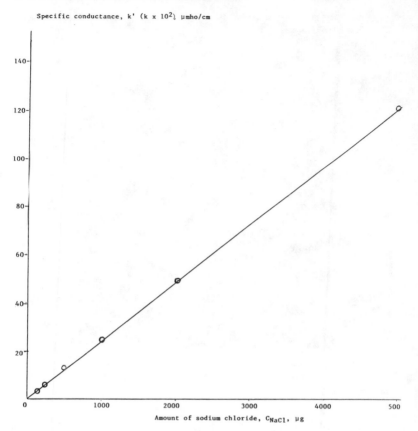

Specific conductance, k' (k x 10^2) μmho/cm

Amount of sodium chloride, C_{NaCl}, μg

Fig. 3.3. Specific conductance versus amount of sodium chloride.

ionic contamination tester was found to be $m = 0.0242$. See Figure 3.3 for a plot of k versus the amount of sodium chloride, C_{NaCl}.

When a given amount of flux is injected into the test tank (see Figure 3.1) the resistivity ρ and hence the specific conductance k will change due to the activator in the flux. With V_{flux} (volume of the flux in microliters) and C_{NaCl} known, a new regression line can be calculated for the flux. If the slope of the regression line for the flux is represented by M, then

$$LV = \lim_{V_{flux} \to \infty} (C_{NaCl}/V_{Flux}) = M/m.$$

Determine M, the slope of the flux regression line, using a procedure similar to

Specific conductance, k' (k x 10²), µmho/cm

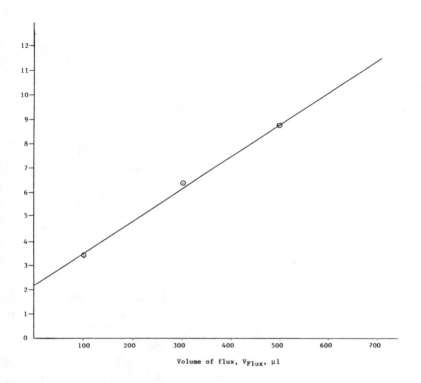

Volume of flux, V_{Flux}, µl

Fig. 3.4. Specific conductance versus volume of flux, RMA#1.

finding m for the sodium chloride line. That is, inject various volumes of flux into the test tank, which is set at two liters, while recording the different specific conductances in the tank.

The slope M is found by plotting the regression line of specific conductances against the volume of flux, measured in microliters. This is presented in Figure 3.4.

Illustration of the Method.   Here is an example of how the procedure was used to find the limit value of one RMA flux. Volumes of 100, 200, and 500 microliters of flux were measured and injected into the test tank set at two liters (2L). These are suggested volumes to be used with RMA fluxes. With fluxes containing more activator, smaller volumes can be used. Several injections of each volume were made and an average resistivity value found.

Figure 3.4 shows the values plotted on the graph. By linear regression, it was found that $M = 0.0134$. The limit value of this flux is

Table 3.2. Limit Values of Twelve Different Fluxes.

PARAMETER	LIMIT VALUE (LV) $\mu g/\mu l$	INCREASING IONIC CONCENTRATION
RMA #1	0.55	
RMA #2	0.81	
RMA #3	0.59	
RA-MIL #1	0.66	
RA-MIL #2	2.90	
RA-MIL #3	4.88	
RA #1	3.15	
RA #2	6.40	
RA #3	20.17	
SA #1	48.06	
SA #2	49.46	
SA #3	49.50	

$$LV = M/m = 0.0134/0.0242 = 0.55 \ \mu g/\mu L.$$

Table 3.2 lists the limit values of twelve different fluxes that were found using this method. It is seen from Table 3.2 that the limit value range of the RMA fluxes is 0.1 to 1.0, for RA-MIL fluxes about 1.0 to 5.0, and for RA fluxes 5.0 and greater. Clearly, RA#3 is a superactivated rosin flux. All of the synthetic activated fluxes have very high limit values. This procedure will hopefully prove useful to anybody who wants to know more about the specific amount of activator in fluxes. It should also prove to be a reliable quality-control measure.

Flux Chemistry. It might be appropriate to say a few more words here about rosin since it is an important ingredient in many fluxes and solder pastes. It is a naturally occurring complex mixture of many different organic acids having the chemical formula $C_{19}H_{(27+0,2,4,6)}COOH$. The number of hydrogen atoms (H) depends on the number of double bonds in the molecule. Naturally occurring rosin also generally contains some inert materials. All rosin comes from pine trees, but there are several sources.[4] If the rosin comes straight from the pine tree in the form of an oleoresinous gum and is subsequently steam distilled, it is known as gum rosin. There are different grades of gum rosin, but water white (WW) gum rosin is generally the grade used in flux manufacture. In spite of its name, water white gum rosin is pale amber in color. Rosin can also come from the papermaking industry as a byproduct of this process. If this is the case, the rosin is known as tall oil rosin. Generally, this product is more consistent chemically than gum rosin. Because of this, tall oil rosin can generally withstand exposure to higher temperatures. This point will be expanded upon later when

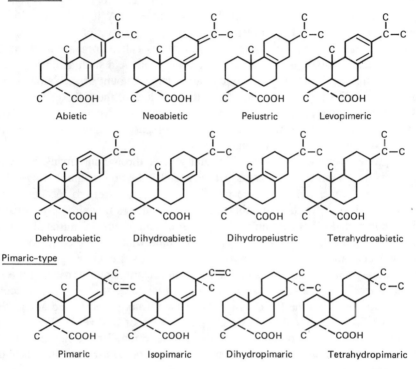

Fig. 3.5.   Structure of the rosin acids.

the subject of white residue (WR) is taken up. Rosin can also be produced from the steam distillation of pine stumps. If this is the case, the rosin is called wood rosin.

Naturally occurring rosin contains a number of different rosin acids, as was mentioned. The principal rosin acids are abietic acid and pimaric acid, and the other rosin acids can be classified as either abietic-type or pimaric-type, depending on their structure.[5] See Figure 3.5 for a depiction of the two-dimensional structural formulas for the principal rosin acids.

Synthetic activated (SA) fluxes, unlike rosin fluxes, are entirely man-made and do not exhibit the chemical inconsistencies of such naturally occurring products as rosin. Synthetic activated fluxes are generally formulated using large organic acid molecules. The molecules chosen are designed to remain liquid and not to solidify so that they can be more easily removed. This is also true of the activators in such fluxes. Since SA fluxes normally contain much higher limit values (see Table 3.2), it is imperative that their residues be thoroughly removed before the completed assembly is put into use. A typical SA flux might consist of an alkyl acid phosphate (vehicle), sulfonic acid (activator) and an alkylamine hydrochloride (activator) in isopropyl alcohol (carrier). SA fluxes are specifically designed to be removed by solvents in the defluxing operation.

Water soluble (WS), or as they are sometimes called, organic acid fluxes, typically use a polyol as the vehicle with an organic acid such as citric acid as the activator. Such fluxes normally contain a large amount of activator, and as in the case of SA fluxes, it is imperative that residues of these fluxes be cleaned as soon as possible after the flux/soldering operation. However, since defluxing involves an aqueous operation rather than a solvent operation, no further discussion of these fluxes will be taken up here.

To promote improved solderability and greater throughput, many PWB assemblers desire more active fluxes. Unfortunately, the greater the fluxing power of the flux in question, the more potential damage it offers to the completed PWB assembly. All of the activating agents used in fluxes are extremely corrosive to the completed PWB assembly, and proper cleaning must be initiated as soon as possible after the flux/soldering operation; otherwise, severe corrosion products will remain on the PWB assembly surface. Fluxes containing chloride ion (Cl^-) are especially deleterious because they are instrumental in causing the so-called corrosion cycle to continue, resulting in white, grainy or powdery compounds appearing on the PWB assembly surface if not properly and adequately cleaned.[6] These white substances are typically salts of lead (Pb). Fluxes which do not contain halides (chloride ion, Cl^-; bromide ion, Br^-) generally contain organic acids such as adipic or succinic as activators. These too can contribute to corrosion problems if not properly removed.

Solder Pastes.   Today surface mount technology (SMT) is much more prevalent, and as was pointed out in Section 3.1.1, Type I and Type II surface mount assemblies (SMAs) have components attached to the substrate by a reflow technique. The chief methods of reflow soldering at present are vapor phase soldering (VPS) and infrared soldering (IRS). For both methods a solder paste rather than a flux is employed.

A solder paste is a material already containing the solder in the form of a solder powder. The powder consists of small spherical or irregular particles of solder. A solder paste also contains rosin (generally in a modified form), activator, and also thickening agents. The thickening agents used are normally derivatives of hydrogenated castor oil which is the triglyceride of ricinoleic acid. These agents are put into the paste to produce a paste having the proper viscosity for screening applications and to control slump. A paste will also contain a solvent carrier, but the solvent is generally a high boiling solvent such as dibutylcarbitol or diethyleneglycolmonobutyl ether [2-(2-butoxyethoxy) ethanol]. These high boiling solvents are put into the paste to control the tack time. So in addition to producing post-solder residues similar to those of fluxes, e.g.,

Table 3.3. Classification of the Chief Contaminant Types on Printed Wiring.[a]

CATEGORY 1*	CATEGORY 2	CATEGORY 3
Resin and fibreglass debris from drilling and/or punching operations	Flux activators	Flux resin
	Activator residues	Flux rosin
	Soldering salts	Oils
Metal and plastic chips from machining and/or trimming operations	Handling soils (sodium and potassium chlorides)	Grease
		Waxes
Dust		Synthetic polymers
	Residual plating salts	
Handling soils		Soldering oils
	Neutralisers	
Lint		Metal oxides
	Ethanolamines	
Insulation		Handling soils
	Surfactants (ionic)	
Hair/skin		Polyglycol degradation byproduct
		Hand creams
		Lubricants
		Silicones
		Surfactants (nonionic)

[a]Contaminants may exhibit characteristics of more than one category.
*Category 1: Particulate
  Category 2: Polar, ionic, or inorganic
  Category 3: Nonpolar, nonionic, or organic
Source: ANSI/IPC-SC-60 1987.

activator and rosin residues, pastes can also produce thickening agent residues and solder balls. The latter must be considered a contaminant since they are an unwanted byproduct of the manufacturing operation that may contribute to unacceptable assembly performance by producing an electrical short.

Summary of Contaminant Types.   Table 3.3 presents a classification of the chief types of contaminants found on printed wiring, both bare boards and assemblies. Chemically these species can be classified as ionic, polar, nonpolar, nonionic. Ionic species are those that in solution form positively and negatively charged species. A good example of this kind of compound is sodium chloride (NaCl) in water. In water sodium chloride forms hydrated sodium ions (Na^+) and hydrated chloride ions (Cl^-). However, a warning here. A compound that forms ions in water may not do so in another solvent, e.g., in trichlorotrifluoroethane (CFC-113) or may do so to a considerably lesser extent.

A polar molecule is by definition one having a dipole moment. This means there exists an unequal distribution of electrons in the molecule which produces the dipole molecule.[7] Some polar species such as hydrogen chloride (HCl) when subjected to various solvents, e.g., water, dissociate to form ions. Again, it depends on the particular molecular species and solvent in question. A nonpolar molecule is by definition one having an zero dipole moment. In addition to taking into consideration the kinds of atoms in the molecule, molecular symmetry must also be taken into account when computing a dipole moment. A nonionic molecule is one that does not form ions (positively and negatively charged species) under any circumstances. Many organic materials are nonionic. The next section will address the effect of different types of contaminants on printed wiring.

3.1.3. Results of Different Types of Contaminants on Printed Wiring

It is clear that potentially a wide variety of contaminants can remain on the completed assembly surface if they are not properly cleaned off. As was pointed out, the contaminants can be ionic, polar, nonpolar, or nonionic. Salts such as plating and etching salts and activators are typically ionic. Many of the organic species are nonionic. However, organic acids such as the rosin acids and organic acids used as activators are ionic. An organic molecule by definition is one based on the element carbon (atomic symbol: C).

The most detrimental type of contaminant is the ionic type. Ionic species if not cleaned off can cause serious problems. Among these are leakage currents between traces and severe corrosion. The presence of moisture greatly accelerates the activity of ionic species since water solvates the ions and enables species since water solvates the ions and enables them to become mobile. In the presence of an applied voltage across circuit traces and moisture, dendritic growth can occur. This phenomenon is where atoms of the circuit trace migrate outward across the bare laminate until they bridge the circuit width causing an electrical short. Mobile ionic species speed up this form of growth. Mobile ionic species are also responsible for corrosion products being formed on the assembly surface.[8,9] It is to protect the assembly surface from moisture that military contractors are required to apply a conformal coating. A conformal coating is, of course, a polymeric material such as an epoxy, polyurethane, or acrylic designed to protect the assembly surface from moisture. In addition, conformal coatings also isolate contaminants and prevent them from migrating or being dislodged during temperature-moisture cycling, high vibration, shock environments, etc.

Nonionic contaminants remaining on the assembly surface can also lead to serious problems. These contaminants can interfere with electrical bed-of-nails testing by creating electrical opens. They can also lead to adhesion problems if a conformal coating is being applied. If this type of contaminant is left on the surface at the bare board stage and a solder mask is being applied, they can lead to adhesion problems with the solder mask. Nonionic contaminants can also encapsulate ionic material which under suitable conditions of temperature and humidity can become mobile species.

Ionic materials and some hygroscopic nonionic materials under conformal coatings can also lead to blistering during temperature/humidity cycling. This phenomenon is also called mealing or vesication, and it takes place because no conformal coating is 100% effective against moisture penetration. In fact, the presence of ionic material or hygroscopic material on the surface under the conformal coating actually acts to promote water penetration since such contamination attracts water penetration since such contamination attracts water molecules. Once the contaminant becomes hydrated, it builds up osmotic pressure which causes a lifting of the conformal coating thus leading to the blister or vesicle. This type of phenomenon must be considered detrimental to the finished assembly.[10]

Finally, there is the phenomenon of white residue (WR). White residue is not always white. It can be gray, tan, beige, or amber in color.[11,12,13] There are several possible causes. In some instances the assembly butter coat is removed, thus revealing the glass weave intersections. This is known as measling. This can result if too harsh a solvent is used for cleaning. Another instance of white residue is when activator materials are left behind on the assembly surface. This can come about if the solvent used for defluxing has become depleted of alcohol (this is especially true of solvents based on trichlorotrifluoroethane or CFC-113). Another cause of white residue is solder paste. Solder paste contains materials known as thickening agents (thixotropic agents or rheological modifiers). These materials are difficult to remove during the defluxing operation, especially without the expenditure of mechanical energy, e.g., sprays. These thickening agent residues are prone to form a white residue, especially if exposed to alcohol. Exposure to alcohol can come about either from alcohol in the defluxing solvent or alcohol in an ionic contamination tester.

However, the principal cause of white residue is due to thermal degradation of rosin.[14] Rosin readily undergoes degradation and polymerization (the molecules bond together to form a much larger molecule). This is specially true when it is heated around 500°F or higher. The rosin also interacts with tin and lead salts formed during the fluxing operation by the activator action on the solder oxides. The rosin interaction products are termed tin and lead abietates. The polymerized rosin and/or tin and lead abietates are much more difficult to remove by a defluxing solvent because of their more insoluble nature. Typically, the white residue appears after the defluxing operation after the defluxing solvent has flashed off. Both rosin fluxes and rosin pastes are prone to form this type of white residue.

If this type of white residue appears, what can be done? First, it is necessary to ascertain that the flux (or paste) is truly the source of the white residue and to verify that the white residue does not have another cause. If the flux (or paste) is the cause of the white residue, one way to prevent its occurrence is to more carefully control the soldering process. Make sure that no more heat than necessary is applied. The thermal profile of the assembly must be carefully set up and controlled. This applies to both wave soldering and reflow soldering. The nonequilibrium conditions taking place in both wave soldering and infrared reflow soldering are especially likely to give this type of problem if rosin-based fluxes and/or pastes are being used. By nonequilibrium conditions are meant that the assembly substrate material is not allowed to reach the temperature of the heat sources, e.g., solder wave, IR lamps, etc., in the soldering process.

Another satisfactory way of dealing with a WR problem is to consider changing fluxes (or pastes). Fluxes based on tall oil rosin are more consistent chemically and thus can be exposed to somewhat higher tempertures than gum rosin fluxes. The latter, because of their nature, are more influenced by the climactic conditions to which the pine trees were exposed and are thus more sensitive to thermal degradation because they are less consistent chemically. Modified rosin fluxes are those where the rosin has been chemically treated to make it more chemically consistent so that it can be exposed to higher temperatures without undergoing thermal degradation. Hydrogenation, or the filling up of double bonds within the rosin acid molecules with hydrogen, is an important way in which rosin is modified to make it more heat-resistant. Most solder pastes today are based on rosin employing a modified rosin to prevent or at least retard the formation of white residue.

If this type of white residue has been formed, it can be very difficult to clean. One suggestion is to expose assemblies with WR to the flux again by passing them over the flux wave either without preheat or possibly some level of preheat and not passing over the solder wave then followed by immediate cleaning. The flux, being of a similar chemical nature to the WR, helps solubilize it. Another technique is to try batch cleaning if the assemblies have WR. Use direct boil immersion in which some flux has been placed in the boil sump followed by vigorous spraying using pure solvent.

Finally, how detrimental is this type of white residue? Certainly it is unseemly and definitely presents a cosmetic problem. Strictly speaking, polymerized rosin may not present a problem per se; however, it is conceivable that it can encapsulate ionic material. The tin and lead abietates, although only very slightly soluble, are polar species, and it is conceivable that certain temperature and humidity conditions may turn them mobile, if ever so slightly. As circuitry lines and traces grow narrower, this could present a problem. So it is probably best to avoid this type of white residue.

3.2. THE CLEANING PROCESS

It is evident from the last section that potentially a large number of contaminants can be left on the PWB assembly surface after processing. The question is: Is cleaning necessary? The answer to this question depends on the assembly configuration: component density, types of components, line/spacing widths and most important—the service conditions under which the assembly will most likely be employed and the degree of reliability demanded by the user.

If the service conditions are likely to be harsh and/or variable and the degree of reliability required is high, cleaning is mandatory. Under such conditions cleaning both the bare board after the fabrication process and prior to assembly operations would be an excellent practice. Defluxing after the flux/soldering operation is mandatory.

Even for assemblies used in less critical environments, it is important to consider what the service environment could potentially be. For example, computers normally operate under controlled temperature/humidity conditions. However, sometimes the air conditioning fails. What happens then? So the question of whether to clean must be asked for each type of assembly. If the consequences of assembly malfunction or failure are not particularly important, cleaning may not be necessary. If the consequences are more serious, cleaning should definitely be considered, and if the consequences are serious or life-threatening, cleaning is mandatory.

The defluxing process can be categorized into several different types today. These are:

1. Solvent cleaning using an organic solvent to dissolve or remove flux and/or paste residues
2. Solvent cleaning followed by an aqueous rinse to produce a clean product
3. Aqueous cleaning using a saponifier to make rosin flux and/or rosin paste residues soluble in water
4. Aqueous cleaning using a water soluble flux and/or paste
5. Semi-aqueous cleaning using a material to dissolve the flux and/or paste residues followed by an aqueous rinse to produce a clean product
6. Use of "no-clean" fluxes—fluxes requiring no cleaning
7. Controlled inert atmosphere soldering machines requiring no cleaning

Only the first two cleaning processes will be discussed here. The others will be taken up in subsequent chapters.

(1)   trichloroethylene

(2)   1,1,1-trichloroethane

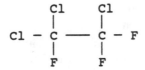

(3)   1,1,2-trichloro-1,2,2-trifluoroethane (CFC-113)

Fig. 3.6.   Structures of (1) trichloroethylene, (2) 1,1,1-trichloroethane, (3) 1,1,2-trichloro-1,2,2-trifluoroethane (CFC-113).

3.2.1. Solvents for Cleaning

Many years ago the solvent of choice in the electronics industry was trichloroethylene. Today it is not very widely used, chiefly because of its toxicity (the issue of solvent toxicity will be addressed below). In the 1960s, 1,1,1-trichloroethane was introduced into the electronics industry. It found wide usage both as a photoresist developer as well as a defluxing solvent. Soon afterwards 1,1,2-trichloro-1,2,2,-trifluoroethane (CFC-113) was introduced. The structures of these three molecules are depicted in Figure 3.6. In the United States and abroad, by far the two most common solvents used for defluxing are 1,1,1-trichloroethane (sometimes known as methyl chloroform) and CFC-113. Small amounts of trichloroethylene and CFC-112 (tetrachlorodifloroethane) are also used as defluxing

Table 3.4. Physical Properties of Pure Solvents Used in Defluxing.

	CFC-113	1,1,1-TRICHLOROETHANE (METHYL CHLOROFORM)	TRICHLORO-ETHYLENE
Molecular weight	187.4	133.4	131.4
Appearance	Colorless liquid	Colorless liquid	Colorless liquid
Boiling point,			
°F	117.6	165.4	188.6
°C	46.6	74.1	87.0
Vapor pressure @ 68°F (20°C),	5.5	2.0	1.1
Liquid density @ 77°F (25°C),			
g/cc	1.565	1.319	1.457
lbs/gal	13.06	10.97	12.10
Kauri-butanol number	31	124	130
Specific Heat @ 68°F (20°C), Btu/lb/°F			
Vapor (saturated)	0.15	—	—
Liquid	0.21	0.26	0.23
Latent heat of vaporization at NBP, Btu/lb	63.1	98.0	103.0
Surface tension @ 68°F (20°C), dynes/cm	18.8	25.9	29.2
Solubility of water @ 68°F (20°C), ppm	90	400	300
Solubility in water @ 68°F (20°), ppm	270	700	1000
Liquid viscosity @ 68°F (20°C), cP	0.69	0.86	0.54
Evaporation rate (carbon tetrachloride = 100)	280	139	69
Flashpoint	None	None	None

solvents. Table 3.4 gives the relevant physical properties of CFC-113, 1,1,1-trichloroethane, and trichloroethylene.

Safety Aspects of Solvents.   Up through the 1950s very little heed was paid to potential health hazards in the workplace in general. This was also true in the electronics industry.

As knowledge of the toxicology of exposure to chemical substances increased, so did concern for the safety of all potentially involved persons, be they in or out of the workplace. The Clean Air Act of 1970, and the Occupational Safety and Health Act promulgated regulations intended to minimize the exposure of humans to potentially dangerous materials. As an offshoot of the Occupational Safety and Health Act (OSHA, hereafter) there came into being NIOSH, the

National Institute of Occupational Safety and Health, which has no regulatory power, but is a recommending body to give guidance to OSHA.

Through the auspices of NIOSH, exposure limits of solvents in the air to which the normally healthy individual could be exposed without bodily harm were set. Of the common solvents in the workplace, the chlorinated solvents such as trichloroethylene and 1,1,1-trichloroethane were decreed to be more harmful than the fluorinated solvents (CFCs). To understand this, one must first understand the rating system and what it means.

There are several acronyms to be defined before full understanding of the hazard potential of various substances can be appreciated. First, one must realize that there are two separate bodies that involve themselves in making recommendations to those governmental agencies charged with promulgating regulations. The first of these, NIOSH, has already been mentioned. The second very important group is ACGHI, The American Conference of Governmental and Industrial Hygienists. These bodies, through their various committee studies, establish the numerical values to be used in rating the relative hazard potential of the chemical compounds found in the workplace. These are given numerical values called TLVs, which stands for Threshold Limit Values. TLVs, by the very nature of their being, are not absolute numbers that define the boundary between perfectly safe and a life-threatening situation. They are an effort to establish an accurate guideline parameter based on the best available experimental data.

TLVs are expressed in three categories, and these are defined as follows:

1. TLV-TWA. This is the threshold limit value–time weighted average. The TWA portion is defined as the time-weighted average concentration of the substance in question to which a healthy workman may be exposed during a normal 8-hour day/40-hour work week, without adverse effect.
2. TLV-STEL. This is the threshold limit value–short term exposure limit. This relates to the concentration to which workmen may be exposed for a short time, provided the TWA is not exceeded. It should not be considered as a separate independent exposure limit, but as a supplement to the TLV-TWA. The STEL is a 15 minute continuous segment of time during which the workman should not be exposed to a concentration higher than the STEL limit, even though the calculated TWA on an 8-hour basis is not exceeded. Also, the STEL must not be repeated more than four times per day.
3. TLC-C. The C in this instance refers to a ceiling concentration which should not be exceeded at any time.

See Table 3.5 for the threshold limit values of the pure solvents used in the electronics industry. For azeotropic solvent systems (generally based on CFC-

Table 3.5. Threshold Limit Values (TLVs) of Solvents Used in Defluxing.

	TWA	STEL
CFC-113	1000	1250
1,1,1-Trichloroethane	350	450
Trichloroethylene	50	200

113) and stabilized solvent blends such as stabilized 1,1,1-trichloroethane, the TLV must be calculated based on the TLVs of the individual components of the azeotrope or blend. The manufacture of the various solvents will supply the calculated TLVs of the solvents.

3.2.2.   Bipolar Solvent Cleaning

When using CFC-113 as a cleaning agent, other organic compounds are generally combined with the CFC-113 to produce a more effective cleaning agent.[15] Typically, lower molecular weight alcohols such as methanol (CH_3OH), ethanol (CH_3CH_2OH), and/or isopropanol ($CH_3CHOHCH_3$) are used to achieve more effective defluxing. The alcohols are both excellent solvents for a wide variety of organic materials and also ionic species. One important solvent based on CFC-113 contains this material and methanol with a small amount of stabilizer. Another CFC-113 solvent, in addition to methanol, also contains small amounts of acetone and isohexane to effectively remove nonpolar species on the PWB assembly surface. There are other solvents based on CFC-113 which contain several alcohols.

The CFC-113 molecule possesses the ability to form azeotropes with a wide variety of materials. It is probably best from the standpoint of machine efficiency to use an azeotrope. An azeotrope, strictly speaking, is a mixture of different ingredients that behaves as if it were a pure compound when it is boiled. That is, it has a single unique boiling point and preserves its composition when it boils and then is subsequently condensed.[16] The formulation based on stabilized CFC-113 and methanol and the one based on stabilized CFC-113, methanol, acetone, and isohexane are azeotropes. There are also other azeotropes based on CFC-113 used in the electronics industry. The physical properties of the solvents will be supplied by their respective manufacturers.

In relation to solvent defluxing, sometimes aqueous rinsing with deionized (DI) water is performed after rinsing; this is called combination defluxing, and its usage was being promoted by the military. In particular, one investigation cited its benefits.[17] Although combination defluxing offers excellent removal of ionics resulting in very low ionic levels, it is costly. In addition to a solvent defluxer, an aqueous rinsing machine with DI water capability is required. In addition to the capital cost, additional floor space for the aqueous rinsing machine is required. As a consequence, combination cleaning is not too frequently used. Solvent defluxing, if the right solvent, equipment, and cleaning cycle are chosen, can result in acceptable cleanliness levels, obviating aqueous rinsing following the defluxing operation.

Solvent systems based on 1,1,1-trichloroethane sometimes employ another organic compound such as an alcohol, but 1,1,1-trichloroethane is a more aggressive solvent than a chlorofluorocarbon (CFC). This is not necessarily advantageous, since it is also more aggressive in its action on PWB substrate materials, components, and component marking inks. In addition, 1,1,1-trichloroethane is spontaneously unstable and typically a solvent based on this material will contain 5 to 8 wt.% stabilizers to prevent 1,1,1-trichloroethane from decomposing into hydrochloric acid. This is in marked contrast to the CFCs, which normally only require a very small amount of stabilizer (<1.0 wt.%). Finding the proper stabilizer package for 1,1,1-trichloroethane has also proved troublesome since one of the most common stabilized compounds used for this purpose, 1,4-dioxane, has been classified as a Class 2B carcinogen, which means it is possibly carcinogenic to humans. However, a toxicology study found no evidence of carcinogenicity for a mixture containing 1,1,1-trichloroethane, 1,4-dioxane, and other ingredients.[18] In addition, 1,1,1-trichloroethane has a much higher boiling point than the CFCs, thus requiring more energy input. However, on the positive side, the higher boiling point of 1,1,1-trichloroethane makes it easier to retain in cleaning equipment.

1,1,1-trichloroethane is an effective solvent for removing rosin residues and its higher boiling point does give it some advantage in this respect because the higher temperature will aid in liquifying solidified rosin residues. Its other most prevalent advantage over the CFC-based solvents is its cost. Typically, solvents based on 1,1,1-trichloroethane are cheaper.

3.2.3.   Equipment for Cleaning

There are essentially two distinct ways of approaching defluxing PWAs and SMAs. The first approach is to deflux assemblies processed in batches. This is known as batch defluxing, and the process equipment in which the defluxing is performed is known at a batch defluxer. In general, batch defluxing is suitable for low volumes/intermediate to high product mix processing operations. Such defluxing is more manually intensive; however, there are now a number of automated hoists that can be used with batch units. A hoist reduces the amount of manual labor involved in the defluxing operation and also removes many process variables. The proper programming of a hoist can also greatly aid in solvent conservation since the hoist can be programmed in such a way that excessive loss of solvent is avoided. A number of different manufacturers offer

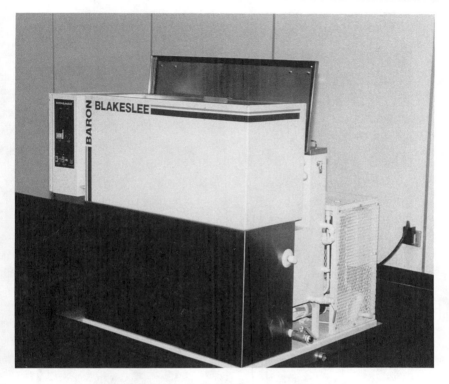

Fig. 3.7. Batch defluxer. (Courtesy of Allied-Signal/Baron-Blakeslee.)

batch units, generally with ultrasonics (US) and a programmable hoist as options. Batch units are normally much more economical to purchase than a conveyorized defluxer and normally require less floor space. See Figure 3.7 showing a typical batch defluxing unit. Figure 3.8 depicts a typical batch defluxing unit with an accompanying programmable hoist.

The second approach is the use of an automated, conveyorized inline defluxer. These machines are much more automated than batch units; consequently they are more suitable for high volume, high throughput operations. Automated material handling devices can be used which transfer the assemblies automatically from the soldering machine (wave solder or reflow) to the defluxer. An inline defluxer has a movable conveyor, generally mesh-belt type, that transports the assemblies from the defluxer entrance to the exit. Because most solvents have adequate flashoff properties, drying often is not necessary. However, because of more emphasis being placed on solvent conservation, drying is now being more frequently considered as part of the defluxing process, especially to reduce workpiece dragout. Workpiece dragout refers, of course, to solvent entrapped in or on assemblies that is carried out of the cleaning equipment and subsequently

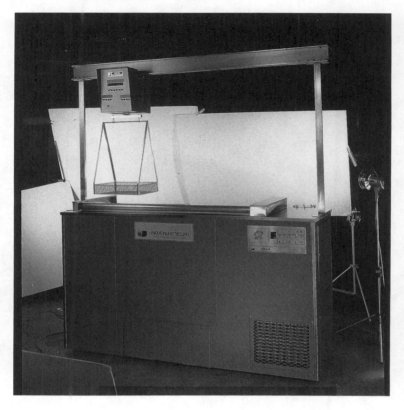

Fig. 3.8. Batch defluxer with programmable hoist. (Courtesy of Unique Industries Inc.)

evaporates into the atmosphere. A conveyorized defluxer also will have several sets of spray headers: a prespray header with moderate spray pressures followed by one or more recirculating spray manifolds at somewhat higher pressures and a distillate header set at low pressures. Immersion in the solvent is also a feature on some inline defluxers.

One approach to an inline defluxer suitable for the lower boiling solvents described in Section 3.4.3 is to use two baffles—one at the entrance and one at the exit—to act as liquid seals. The use of these two baffles acts to isolate the high pressure recirculating spray manifolds from the air, thus preventing large solvent vapor losses. Another solvent vapor conservation technique is the use of superchilled auxiliary coils (kept as $-25°F$) located above the primary vapor condensing coils. These superchilled auxiliary coils help keep a cold blanket of air over the solvent vapors, thus aiding greatly in solvent conservation. See Figure 3.9 showing a typical conveyorized defluxing unit. Figure 3.10 is a schematic diagram of this equipment. A machine design that conserves solvent

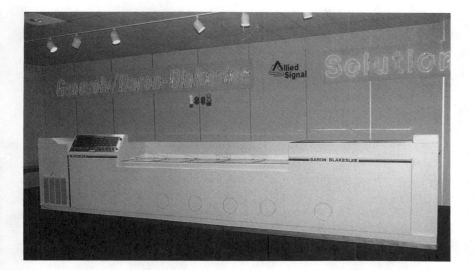

Fig. 3.9.   Conveyorized defluxer. (Courtesy of Allied-Signal/Baron-Blakeslee.)

usage will definitely be a requirement for new solvent-handling equipment. Other techniques for vapor retention may also be feasible.

3.2.4.   The Defluxing Process

Both types of defluxing equipment have now been discussed. If a batch unit is used, for effective defluxing a minimum of two sumps in the batch unit along with distillate spray capability is highly recommended. One sump will be the boil (vapor) sump; the second sump, the rinse sump, holds cleaner solvent. As the solvent vapors boil and are condensed at the primary condensing coils, clean solvent (condensate) is returned to the rinse sump which overflows into the boil sump.

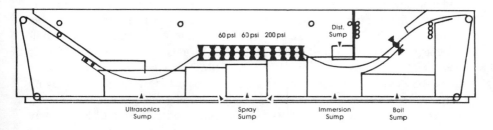

Fig. 3.10.   Schematic of conveyorized defluxer with liquid seals. (Courtesy of Allied-Signal/Baron-Blakeslee.)

For effective cleaning of PWAs with conventional throughhole components (THCs), an adequate cleaning cycle would involve vapor immersion over the boil sump until vapor condensation on the PWAs ceases. This may then be followed by spraying (below the vapor line) the assemblies. Most of the contamination will end up in the boil sump. The assemblies can then be immersed in the rinse sump and then placed in the vapors again. Spraying may again be done, after which the assemblies are allowed to remain in the vapors until vapor condensation ceases. They are then slowly withdrawn from the machine. If a solvent recovery still is used in conjunction with the batch unit, a very effective method of defluxing is to directly immerse the assemblies in the boil sump solvent, provided the boards and components can stand the temperature. No more than a 30 second immersion should be necessary. If a solvent still is not being used, direct boil immersion is not recommended due to contaminants in the boil sump recontaminating the assemblies.

Many batch units have ultrasonic (US) as an option. If US cleaning is employed, it is generally performed in the rinse sump. Regarding the use of US for cleaning PWB assemblies, this is still a moot point. The question does not concern cleaning effectiveness because there is no doubt that US greatly enhances the cleaning efficacy of solvents and cleaning solutions. US cleaning is especially effective in crevice cleaning, allowing the solvent to clean in tight toleranced and narrow dimensional parts of the assembly. It is still an open question, however, whether US damages the internal wiring bonding of packaged integrated circuits (ICs) on the board surface. Earlier US work performed at 20 kHz frequency did lead to internal IC wire bond damage. However, today frequencies of 40 kHz and higher are common, and there is some evidence that higher frequencies are not as harmful as lower ones. The Electronics Manufacturing Productivity Facility (EMPF)/Naval Weapons Center in Ridgecrest, California has an active program investigating this important issue.

One more point about batch cleaning. Although PWAs with conventional THCs have been satisfactorily defluxed in batch units, surface mount assemblies present a much more difficult challenge because of the smaller standoffs of SMCs and generally tighter pitches. The pitch of a component is defined as the center-to-center distance between two adjacent leads of a component. For throughhole components (THCs), the pitch is typically 100 mils, for surface mount components (SMCs), the typical pitch is probably 50 mils, but SMCs with 25 mils and even less are now common. With leaded SMCs, a vigorous cleaning process involving direct boil immersion and spraying would have to be used. Spraying may not be too effective because batch units generally have low pressure spray wands. If some paste residue is acceptable, batch unit defluxing of SMAs may be satisfactory, especially if the SMCs on the SMA are leaded. Generally though, for SMAs conveyorized inline defluxing is the preferred method.

Both PWAs with conventional THCs and all three types of SMAs can be defluxed in an inline machine. For cleaning PWAs, low to moderate pressures (10–30 psi) are generally sufficient. For SMAs higher pressures (50 psi $< P <$ 200 psi) are generally required. This is because of the smaller standoffs of the SMCs and also because paste residues are much more tenacious and difficult to remove effectively. In addition to higher pressures, a fairly high volume of solvent also seems necessary. Some research suggests that spray angles and spray types may also play a role in cleaning whereas other pieces of research suggest that these factors are not particularly important.[19] The importance of these parameters regarding their effect on cleaning is still under investigation.

3.2.5. Solvent Conservation and Reclamation

Mention has been made already about conserving solvent while operating a machine. Obviously, with any machine using solvent the machine should be in good working condition. When using a batch unit, good solvent conservation practices include:

1. Make sure the PWAs (or SMAs) are racked properly to give adequate drainage to avoid excessive dragout.
2. The basket in which the PWAs (or SMAs) are placed should be sufficiently smaller than the sump; otherwise, solvent vapors may be forced out by the so-called piston effect.
3. The basket in which the PWAs (or SMAs) are placed should be lowered and raised uniformly at not too great a speed to avoid pumping or dragging solvent vapors out of the machine—an automated hoist is highly recommended.
4. Use a solvent recovery still to purify solvent for machine use and to capture unwanted residues.
5. If spraying is used, make sure all spraying is performed well below the vapor line.
6. Avoid having drafts in the area since they will unduly disturb the vapor blanket and lead to solvent losses.
7. Use a machine with at least 100% freeboard.
8. Use a carbon adsorption system to capture escaping solvent vapors.
9. Train operators in the proper operating procedures and make sure discipline is maintained to ensure that proper procedures are being followed.

For an inline machine, good solvent conservation practices include:

1. Make sure that PWAs and/or SMAs are designed properly to avoid excessive dragout since the conveyor is horizontal in the heart of the machine.

2. If the PWAs and/or SMAs are palletized, make sure the pallet is designed properly to minimize dragout.
3. Use a solvent recovery still.
4. A liquid seal machine with an auxiliary vapor trap is an excellent way to prevent unnecessary solvent losses.
5. Avoid having drafts in the area.
6. Train operators in proper operating procedures.

In addition to solvent conservation practices, solvent reclamation is also a wise and economical practice. This can be accomplished in house using a solvent recovery still in conjunction with the defluxing machine. In general, a solvent still is advisable since it keeps the solvent in the machine relatively pure and captures unwanted residues. Keeping the solvent in the machine pure, even in the boil sump, will allow direct boil immersion cleaning, a very effective cleaning method. If a solvent recovery still is not used, a good solvent reclamation program should be instituted. Even if the user does not wish to use reclaimed solvent, the used solvent in the facility can still be sold to reclaimers. So it is well worthwhile to follow this practice.

3.3. TESTING FOR CLEANLINESS

Today there are a few industry-accepted techniques for determining cleanliness levels after the PWB assembly has been defluxed. However, a great deal of work needs to be done to promote other techniques that have been shown to be effective in ascertaining contamination levels. In addition, work still remains to be performed to correlate the kind and degree of contamination with a particular kind of degradation of the PWB assembly's performance.

The trend in PWB assemblies is towards circuitry with traces and spacings becoming smaller ($<$ 5 mils), fine pitch components ($\leq$ 25 mils), and smaller component standoffs. All of these features will increase the difficulty of achieving adequate cleanliness levels. Concomitant with this will be the need to use more sophisticated techniques and methods for determining if desired cleanliness levels have been achieved. In this section, older methods and also more recent developments will be discussed.

3.3.1. Overview of Contamination Testing

Cleanliness testing started in the early 1970s with the pioneering work of T.F. Egan at the bell Laboratory. Egan employed a conductivity meter and made aqueous extracts to investigate the effect that plating salt residues had on specific conductance.[20] By this time, assembly component densities and line widths and spacings were becoming small enough ($\sim$15–10 mils) that the detrimental effects of contaminant residues were beginning to be noticed. The first test for contaminant residues to be developed was for ionic residues using a simple conductivity cell to measure the specific conductance of a test extract. R.J. DeNoon and W.T. Hobson at the Navy Avionics Center (NAC—however, at that time it was known as the Navy Avionics Facility) determined that a text extract employing 75 vol.% isopropanol (IPA) and 2.5 vol.% water (H_2O) was the most suitable for dissolving rosin-based flux residues following by testing with a conductivity meter.[21] This test solution for ionics has remained standard up to the present day.

This cleanliness test became incorporated in MIL-P-28809, the DoD speci-fication dealing with the acceptability of military PWB assemblies. Soon after this, several companies came out with a more automated version of performing this test since the manual method is tedious and must be performed with extreme care. Today these automated machines have programs stored in EPROMS so that the operator merely has to enter the dimensions of the assembly and adjust the volume of the test solution.

Although the solvent extract resistivity test, or as it is sometimes called, the ionic contamination test, proved relatively easy to implement as a way of de-termining industry cleanliness of PWB assemblies, not much progress has been made for arriving at as simple a test for other types of contaminants that is acceptable by most members of the industry. The ionic contamination test itself suffers from several drawbacks. It does not distinguish among different ionic species, and it does not detect the presence of nonionic and nonionizable (but polar) species even if such species are soluble in the test medium.

Some work has been performed using direct surface analysis techniques (ESCA, Auger, SIMS, etc.), but these methods are generally too costly, both in terms of equipment and personnel, to be used on a routine basis. The results gleaned from these methods are also not entirely conclusive.[22] Because of this, extractive techniques have proved to be the methods of choice for cleanliness testing. A great deal of progress has been made in employing extractive techniques for cleanliness testing. However, what is required is an industry consensus that one or several of these techniques is of direct benefit, and its subsequent incorporation into an acceptable industry standard.

There are several nonextractive techniques that are used by industry members. These methods are based on the application of a bias voltage across a special trace pattern while the PWB assembly is subjected to either moisture or mois-ture/temperature cycling. Surface insulation resistance (SIR) testing and elec-tromigration (EM) testing fall under this category. These techniques will be discussed in more detail below.

3.2.2. Testing for Ionics

Mention has already been made of the solvent extract resistivity test or so-called ionic contamination test. As was pointed out, this test has been around for many years in the electronics industry, and the equipment available is semiautomated and easy to use. See Figure 3.1 (p. 72) showing a standard ionic contamination tester. This test has come under increasing scrutiny because of the impact of surface mount technology (SMT). The chief issue in question is whether the cleanliness test solution (75 vol.% IPA/25 vol.% H_2O) will penetrate under the smaller standoff (2–8 mils) SM components which are finding more frequent usage on assemblies. Because of this issue, ionic contamination test equipment manufacturers have introduced equipment having spray headers in the test tank and the capability of heating the test solution to aid it in penetrating under small standoff components. Figure 3.2 p. 73 shows an ionic contamination tester having these newer features. There are data available to indicate that this approach is sufficient to keep the ionic contamination test as a viable technique for testing cleanliness when surface mount assemblies are being tested.[23]

Ion Chromatography (IC).   Although the ionic contamination test fulfills a definite industry need, it does not and can not distinguish among different ionic species. One may opine that this is not necessary. However, the capability to distinguish among different ionic species and to quantify the amount of each would result in much more effective process control. It has been demonstrated that different ionic species come from different parts of the fabrication and assembly operations, and the ability to distinguish different species would enhance one's ability to control the entire operation more effectively.

To detect and quantify the amount of each ionic species a different analytical technique must be used. This technique is called ion chromatography (IC). As in all chromatographic methods, columns are used to separate a given mixture so that the constituents exit at different times (the elution time), and a detector is used to detect the species in question. The detector signal is converted into an analog signal which appears as a curve on graph paper. The area under this curve can be related to a corresponding area of a standard and in this way the amount of the species in question can be quantified. Each species in question has its own characteristic elution time. Today many chromatographs also have digital readouts, that is, the detector signal is converted into a direct digital signal. It is appropriate here to discuss ion chromatography in more detail.

First, the chemical-suppression approach to ion chromatography will be described. In this approach, a liquid solution called an eluent is mixed thoroughly with the sample solution in question. A suitable pumping system mixes them and drives the liquid through the system. Consider the case of anion analysis. One has a sample, let us say an extract from a printed wiring assembly, and the object is to profile the different ionic species in the sample. By profiling is meant to both detect the species in the sample and to quantify how much of each is contained in the sample. Anions are ionic species having a net negative charge, such as chloride, Cl^-; bromide, Br^-; sulfate, SO_4^{2-}, etc. The separator column must be ion exchange resin having the same anionic end as the eluent anion. As the sample material and eluent pass down the column, sample ions exchange themselves with ions on the column. Each type of ion in the sample is retained for a different time on the column; hence, each type of ions elutes from the column at a different time, the time being characteristic of the type of ion in question.

For example, if sodium hydrogen carbonate ($NaHCO_3$) is the eluent, then the separator column will be of the form: $Resin\text{-}HCO_3^-$. The ion exchange between sample species and the column is:

$$Resin\text{-}HCO_3^- + M^+X^-(sample) \rightarrow Resin\text{-}X^- + M^+HCO_3^-$$

As pointed out, the retention time of the anion species X^- on the separator resin column is characteristic of that species. The stronger the affinity of the species for the resin in the column, the longer the elution time. Hence, separation of the anionic species in the sample occurs in the separator column. The detector is an electrical conductivity cell.

However, if only a separator column and a conductivity cell were used, the eluent would swamp the conductivity cell since the eluent too is ionic in nature. Hence, an additional column called a suppressor column must be employed. The purpose of the suppressor column is to tie up the eluent ions to prevent them from reaching the conductivity cell detector. The suppressor column is made up of a resin that neutralizes the eluent. For anion analysis this will be a strong resin acid in hydrogen form. Two reactions occur in the suppressor column: (1) the eluent ions are neutralized and (2) the sample cations are exchanged for hydrogen ions. These reactions can be represented as follows:

(1) $Resin\text{-}H^+ + NA^+HCO_3^- \rightarrow Resin\text{-}Na^+ + H_2O + CO_2$
(2) $Resin\text{-}H^+ + M^+X^- (sample) \rightarrow Resin\text{-}M^+ + H^+X^-.$

Unlike the ion exchange in the separator column, the exchange in the suppressor column is permanent. Hence suppressor columns can become saturated and must then be regenerated. In more recent applications suppressor columns have been replaced by membrane suppressors.

The different anionic species in acid form ($H^+X_1^-, H^+X_2^-$, etc.) are each eluted at different times and detected by the conductivity cell. As pointed out, the use of a standard having known quantities of each species enables one to quantify the amount of ion in the sample by comparing the area under the curve of the ion in the sample to the corresponding area under the curve of the same ion in the standard.

Cations such as sodium, Na^+; potassium, K^+; calcium, Ca^{2+}, etc. can also be detected and quantified using the same principles as given above for anionic determination, but the separator and suppressor columns must of necessity be

Fig. 3.11. Ion Chromatograph. (Courtesy of Dionex Corporation.)

different from those used in anion analysis. Using IC techniques, it is also possible to detect complex ionic species such as amine hydrohalides, $R_1R_2R_3NH^+X^-$. This is of considerable interest since materials such as these are used as activators in many fluxes. See Figure 3.11 depicting a chemical-suppression ion chromatograph.

One serious disadvantage of chemical-suppression ion chromatography in the past was the necessity of using only aqueous samples. Materials such as rosin flux residues would first have to be extracted in an organic material insoluble in water and then an aqueous extract made. However, more recent developments in ion chromatography have opened up new vistas as to what can be accomplished by this technology. Recently, new types of columns and new developments in instrument materials of construction, e.g., PEEK (polyetheretherketone), make it now possible to analyze for a wide variety of ionic species in both aqueous and nonaqueous samples. Another approach to IC is the use of only a single column (the separator column). Since only a single column is used, this IC technology is known as single-column ion chromatography, or SCIC.[24] Because of these more recent developments in both chemical-suppression ion chromatography and single column ion chromatography, it seems likely that in the future ion chromatographic techniques will play an increasingly important role in the detection and quantification of ionic species on the surface of electronic devices.

The feasibility of using ion chromatographic techniques for detecting and quantifying ionic species in extracts made from printed wiring has been demonstrated.[25,26] In addition, a system capable of detecting both a wide variety of metallic ionic species and organic ionic species has been shown to be feasible.[27] In the future as technology is driven towards even greater miniaturization, such methods may prove very fruitful.

3.3.3.   Testing for Nonionics

As was pointed out in Section 3.1.3, nonionic species can also lead to problems with processing the PWB assembly or interfere with its intended function. To cite one prominent example, in 1980 a major aerospace firm noted that test PWB assemblies after being conformally coated failed preliminary environmental testing per MIL-STD-810C. Those same PWB assemblies, prior to being conformally coated, passed ionic contamination testing. Subsequent characterization of the contaminating species showed it to be a nonionic alkylphenylglycol ether, but its source could not be exactly pinpointed.[28]

A suitable cleaning methodology was devised to clean the remainder of the PWB assemblies produced which had not been conformally coated in order to salvage them. This methodology specified the use of several chlorofluorocarbon solvent systems followed by a 2-propanol rinse. The entire methodology was specifically designed to remove the nonionic glycolether and some other contaminants. Test PWB assemblies were taken after the cleaning protocol had been used, and they subsequently passed environmental testing per MIL-STD-810C.

This whole episode brought vividly to light that no adequate method existed to detect and quantify nonionic contaminants on PWB assembly surfaces. A qualitative method based on infrared (IR) spectrophotometry was developed which has proved useful for identifying organic compounds based on their IR spectrum. It is an extractive technique using acetonitrile as the extractive medium. It is now an IPC (Institute of Interconnecting and Packaging Electronic Circuitry) standard test method, namely, IPC Test Method 2.3.39. The technique involves slowly dripping 0.25–0.50 milliliters (mL) of acetonitrile onto the test specimen (the FR-4 coupon), allowing it to wash across the surface and then to drop onto an ATR plate for insertion into an IR spectrophotometer with a special ATR cell attached. The technique is known as attenuated total reflectance (ATR) or multiple internal reflectance (MIR) spectrophotometry. In this method, one was basically taking the IR spectrum of the organic material on the epoxy-fiberglass surface extracted by the acetonitrile.[29] It is a qualitative method only. However, the method is presently being developed into a quantitative technique.

High Performance Liquid Chromatography (HPLC).   Another method that has been recently developed to detect nonionic species on PWAs and SMAs is liquid chromatography, sometimes known as high performance liquid chromatography or HPLC. Unlike the above method, the method based on HPLC will both detect and quantify the amount of residue. In addition, if peak identification

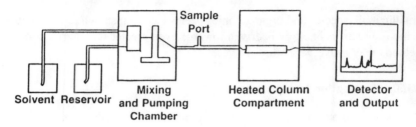

Fig. 3.12.   Schematic of high performance liquid chromatograph.

can be made HPLC is species specific, i.e., it both detects and quantifies the amount of single species such as abietic acid.

Like any chromatography system, it involves the use of an extractive medium (generally acetonitrile), an injection system, columns on which different materials are selectively absorbed so that they elute out of the column at different times, and some sort of detection device. In the case of nonionic residues derived from rosin fluxes and pastes, the most suitable detector is a UV/VIS (ultraviolet/visible) detector which senses any species absorbing light in the ultraviolet or visible range of the spectrum (180 nm to about 700 nm). Figure 3.12 shows a schematic illustration of a liquid chromatograph. Figure 3.13 shows a typical chromatogram

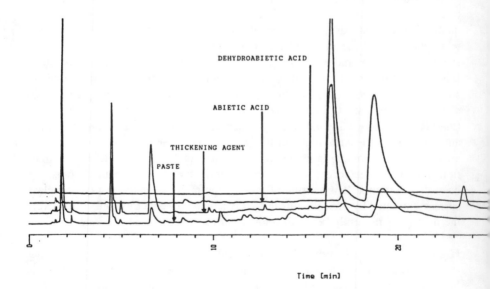

Fig. 3.13.   HPLC chromatogram showing peaks for abietic acid, dehydroabietic acid, and thickening agents.

produced by an HPLC showing peaks for abietic acid, dehydroabietic acid, and thickening agents, the three most prominent ingredients in residue from rosin-based solder pastes. In addition to proving beneficial in detecting contaminant residues, HPLC has also proved useful in setting process parameters for the defluxing operation.[30] Although the HPLC technique is nondestructive, the equipment needed is relatively expensive.

Use of HPLC to Quantify White Residue.   Mention has already been made of white residue (WR). Of all the existing techniques described above, the most promising one for analyzing WR resulting from thermal degradation flux residues is liquid chromatography.[31] As was pointed out, this method is both qualitative and quantitative and has the advantage of being species specific, provided standards are used by which identification on a chromatogram can be made. Several investigations of WR from flux degradation using HPLC strongly implicate metal abietates as the chief promoters of WR on PWB assembly surfaces. The metal abietates are chiefly those of tin and lead. An abietate refers to the salt formed between a metal and any rosin acid (see Figure 3.5). There are also additional data existing implicating tin/lead abietates as the chief components of this type of WR.

The formation of tin/lead abietates undoubtedly takes place during the flux/soldering process. During the time when the activator acts on the tin/lead oxides, tin/lead salts of both rosin and chloride (Cl^-) ion are probably formed. These are termed monoabietate metal salts. Due to excessive heat and in the presence of moisture, these salts dimerize (the molecule expands to approximately twice its size) to form diabietate and didehydroabietate metal salts, both of which are much less soluble than the monoabietate metal salts. These diabietate and didehydroabietate metal salts are formed during the soldering process; however, the white residue normally appears after the defluxing process since many solvents do not possess sufficient solvating power to remove these tenacious species.

Ultraviolet/Visible Spectrophotometry.   Ultraviolet/visible spectrophotometry is also a useful technique for detecting and quantifying contaminant species.[32] Extractions from printed wiring are made using isopropanol (reagent grade or better). Extraction tanks made from stainless steel are probably the most suitable for this purpose, but plastic bags have been used. It is important to calibrate the instrument using a flux to produce the calibration curve. It is best to calibrate using the actual flux or paste whose residues are being determined. The rosin acid residues typically absorb between 200–250 nanometers (nm). Values of rosin residual typically run between 200 and 600 micrograms per square inch ($\mu g/in.^2$). The test is nondestructive.

Coulometry.   Coulometry is another method that has proven successful in de-

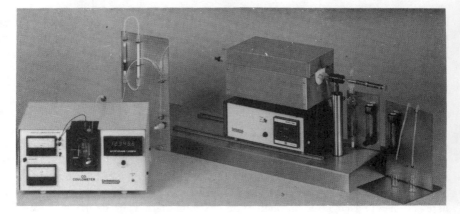

Fig. 3.14. Coulemeter. (Courtesy of UIC Inc.)

tecting organic residues.[33] The method is rapid, simple, reproducible, and quantitative. Coulometry involves the combustion of sample carbon and its conversion into carbon dioxide (CO_2) inside a tube furnace held at 400–500°C. Once the sample carbon is converted into carbon dioxide and dried, the carbon dioxide is reacted with ethanolamine to form carbamic acid [$HOCH_2CH_2NHCOOH$], and the carbamic acid is titrated coulometrically. Since this method essentially involves the pyrolysis of organic material and its conversion into carbon dioxide, it cannot be used with plastic components or with laminate materials.

This method is applicable with ceramic carriers; however, the method quantifies only the amount of carbonaceous residue remaining on the ceramic component and therefore is not species specific. Since the ceramic components must be removed to perform the test, the test is by necessity destructive. It is very effective, however, in quantifying the amount of organic residue left on ceramic components and can be used as a process control aid and process parameter aid. Results are typically expressed in micrograms of carbonaceous material per square inch (μg C/in.^{2}). Figure 3.14 shows the coulometer system.

Turbidimetry. Turbidimetry is also a method that can be used to detect and quantify nonionic (principally rosin) residues on printed wiring.[34,35] It is an extractive technique using isopropanol (reagent grade or better). The turbidimeter must be calibrated using standard solutions. Once extracts are made, hydrochloride acid (0.5N) is added to it to turn it cloudy. The more cloudy the extract, the more rosin residue it contains. The results are given in micrograms of rosin per square inch (μg rosin/in.^{2}).

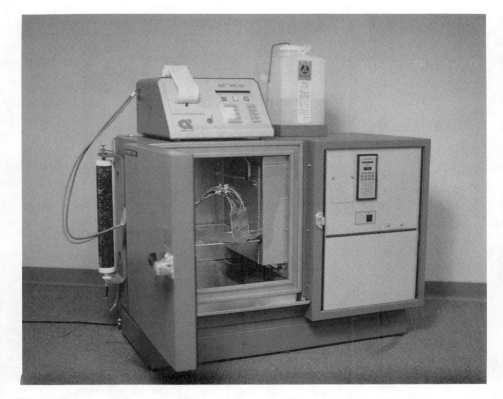

Fig. 3.15.   Setup for performing SIR measurements. (Courtesy of Alpha Metals Inc.)

3.3.4.   Surface Insulation Resistance (SIR) Test

Among the few nonextractive test methods for ascertaining cleanliness of printed wiring, surface insulation resistance (SIR) testing is probably the most important. Chiefly, it involves determining the resistance between parallel traces of a Y-pattern or comb pattern on bare P/I structure laminate material. Such a resistance will normally be very high (in the order of 10^{11}–10^{12} ohms) unless something on the surface causes it to drop. Thus the test can function as a method for checking for contamination left on the surface.[36–43]

The test is typically run at elevated temperatures and humidity over an extended period of time. Figure 3.15 shows a typical setup for measuring SIR. The test patterns used are normally exposed to a bias voltage (excitation voltage) followed by a measurement voltage when actual measurements are being made. The excitation voltage and the measurement voltage may be the same polarity or they may be different. Certain types of contamination under the conditions of

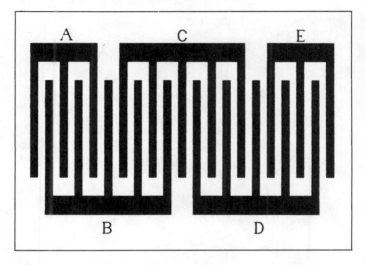

Fig. 3.16. Typical SIR test pattern.

elevated temperature and humidity and excitation voltage can lead to leakage currents across the conductor traces of the test pattern leading to a lowering of the surface insulation resistance.

At present there is no one industry accepted test pattern for SIR testing. Examples of such test patterns are the three found in the IPC-B-25 test board and the BELLCORE test pattern. Figure 3.16 shows a generic SIR pattern. To relate different test patterns to each other, it is important to express all resistance readings in ohms per square. Table 3.6 shows a comparison of various SIR

Table 3.6. Comparison of Various SIR Test Patterns.

PATTERN TYPE	NUMBER OF SPACES	LENGTH OF LINE, INCHES	SPACE WIDTH, INCHES	NUMBER OF SQUARES	MINIMUM RESISTANCE VALUE, $\times 10^9$
One square	1	1.0	1.0	1.0	337.0
Bellcore*	5	1.125	0.050	112.5	3.0
IPC 100043, J1	17	1.0	0.020	850.0	0.4
IPC 100043, J2	17	1.0	0.010	1,700.0	0.02
IPC B-25, A	23	0.625	0.00625	2,300.0	0.015
IPC B-25, B	11	0.625	0.0125	550.0	0.61
IPC B-25, C	5	0.625	0.025	125.0	2.7

*This is the pattern used to meet BELLCORE requirement. All other values in the last column are related to this value.
Source: D.I. Willson, " The Trials and Tribulations of Implementing Surface Insulation Resistance Testing."

patterns. It will be noted that the number of squares is found by dividing the length of the line in inches by the spacing width in inches and then multiplying by the number of spaces.

Nevertheless, even if SIR data are reported in ohms per square, a standard test pattern and comparable excitation and measurement voltages should be used. For example, there is typically a large difference in ohms per square values for a 0.025"/0.025" comb pattern versus a 0.006"/0.006" comb pattern or between measurements made at 10 volts versus measurements made at 100 or 500 volts.

This lack of any standard pattern and standard voltages is one drawback of SIR. Another is the length of time it takes to perform the test. A seven to ten day test is typical and some SIR tests are even extended for six weeks. Finally, although SIR testing will reveal the presence of contamination that causes leakage currents, it is not species specific. It can not, by itself, inform one if the contamination is ionic or nonionic, e.g., a polyglycol.

3.3.5. Electromigration (EM) Test

Another test closely connected with SIR is electromigration (EM) testing.[44] This test involves placing a bias voltage across two conductor traces and determining the time to failure. The time to failure is generally determined by picking a threshold value for the current, e.g., 500 microamps. The time from when the test is initiated to reach this leakage current is then considered the time to failure. Some type of moisture exposure is required to conduct the test. Under the proper bias voltage, dendritic growth takes place between the traces causing a rise in current and a breakdown in the resistance between the traces. Typically a bias voltage less than 10 volts is best for producing dendritic growth. Some people run the test in a temperature/humidity chamber with the same pattern used for SIR testing; some use a drop of deionized water under ambient conditions to produce the phenomenon.

The EM test does indicate the presence of contamination on the surface that can lead to leakage currents and dendritic growth. Also, if the proper test pattern and bias voltage are used, the test can be performed quickly. The chief drawback of the test is lack of any standardization. At present there is no standard test pattern, no standard bias voltage, and no standard methodology for performing the test. Also, the method is not species-specific. It would be extremely helpful if one (or more) standard methods existed for this useful test.

3.4. ENVIRONMENTAL CONCERNS OF SOLVENTS

Solvents, both those based on CFC-113 and those based on 1,1,1-trichloroethane, have experienced wide usage in the electronics industry. Solvent cleaning and defluxing has proven beneficial. The equipment is relatively simple and easy to operate, and solvents have outstanding properties for removing contaminants from the surfaces of PWAs and SMAs. In particular, solvents formulated using CFC-113 have many excellent properties. Chief among these are its outstanding stability and compatibility with almost all materials used in the electronics industry, low toxicity (i.e., high TLV, for example, $TLV_{CFC-113} = 1000$), non-inflammability, and ability to azeotrope with other ingredients—especially lower molecular weight alcohols and other select ingredients—thus enhancing its solvency power for contaminant residues.

3.4.1.   The Ozone Depletion Problem

There is, however, accumulating scientific evidence that chlorofluorocarbons (CFCs) are damaging to the layer of ozone (O_3) in the stratosphere.[45] In fact, any chlorine (Cl) containing molecule, such as 1,1,1-trichloroethane (CCl_3CH_3) or 1,1,2-trichloro-1,2,2,-trifluoroethane (CCl_2FCClF_2)=CFC-113, has the potential for destroying ozone if it reaches the stratosphere. The stratosphere is, of course, the ozone-rich layer of the earth's atmosphere beginning roughly six to ten miles above the earth's surface and extending to about a height of thirty miles. In March of 1988, a distinguished panel of world atmospheric scientists concluded that the amount of total column ozone in the Northern hemisphere had been depleted by 1.7 to 3.0% since 1969, and that man-made chlorine species were strongly suggested to be a chief contributor to this decline. Undoubtedly the excellent stability of CFCs contributes to this problem since they are not subject to breakdown until they reach the stratosphere.

Atmospheric chemistry processes are rather involved. There is some evidence that heterogeneous chemical reactions may be responsible for ozone depletion problems over the polar regions (primarily Antarctica) during the unique conditions of the polar winter. This means that the chiefly destructive reactions do not occur in one phase, for example, the gaseous phase, but in several phases. In fact, particles in the stratosphere are believed to provide surfaces on which these chemical reactions take place involving the destruction of ozone.[46,47]

It is almost certain that active chlorine is the main culprit. It propagates a free radical chain reaction whereby one active chlorine-containing fragment catalyzes the destruction of about one hundred thousand ozone molecules before the reaction is finally terminated. In the case of the Antarctic ozone hole, it appears that the chlorine, which may come initially from man-made chlorine-containing species, such as CFCs and also 1,1,1-trichloroethane (CH_3CCl_3), is decomposed into inactive reservoir species such as hydrochloric acid (HCl), and chlorine nitrate ($ClONO_2$). Laboratory studies have shown that polar stratospheric clouds (PSCs) convert photochemically inactive species such as HCl and $ClONO_2$ into active forms such as chlorine monoxide (ClO) which are capable of destroying ozone molecules. When the temperature drops below a certain point, these reservoir species are entrapped in frozen cloud particles. In the Antarctic spring the reservoir compounds start vigorously releasing active chlorine which then attacks the ozone. The impact of the heterogeneous processes on both polar region and global ozone diminution is not fully understood at this time. There is much that must be learned regarding this critical aspect of atmospheric chemistry.

The Montreal Protocol.   In recent years the ozone problem has been broached under the auspices of the United Nations. In 1987, the so-called Montreal Protocol was proposed by the United Nations Environment Program (UNEP) to regulate CFCs (and halons).[48] The Montreal Protocol is expected to be ratified by a majority of signatory nations. Basically, it says that CFC consumption (consumption = production + imports − exports) will be fixed at 1986 levels by 1989. In 1993 the consumption of CFCs will be phased down to 80% of the 1986 level and by 1998 consumption will be cut back to 50% of the 1986 level. However, based on further scientific findings, which will be the basis of advanced notice of proposed rule making (ANPRM), it is highly likely that the consumption levels may be cut back even more.

The CFC materials covered by the Protocol are: CFC-11, -12, -113, -114, and -115. In addition, Halon-1211, -1301, and -2402 are covered. The halon molecules contain bromine in addition to chlorine and fluorine. The bromine radical is even more destructive towards ozone than chlorine. For example, the ozone depletion potential (ODP) of Halon-1301 is 10.0 as opposed to CFC-11 which is taken as a reference standard having an ODP of 1.0. The ODP of CFC-113 is 0.8. CFC-11 and CFC-12 make up about 75% of the CFCs regulated by the Protocol. For the most part CFC-11 finds use as a foam blowing agent and CFC-12 as a refrigerant.

Although 1,1,1-trichloroethane was not included in the Montreal Protocol as it was originally formulated, the 1,1,1-trichloroethane (methyl chloroform) molecule also depletes ozone, and therefore it will undoubtedly be included as a potential ozone depleting substance by regulatory bodies such as the EPA. At the present time the ODP of 1,1,1-trichloroethane has been set at 0.12.

Stratospherically Safe Materials.   Because of the problem of ozone depletion and the Montreal Protocol, new materials are being actively sought to replace the old solvents. The search for new materials has been especially intense in the case of finding a substitute for CFC-113.

Since these new materials are much less detrimental to stratospheric ozone, they are generally known as stratospherically safe fluorocarbons (SSFs). The principal kind of solvent molecules to emerge as SSFs are the hydrochlorofluorocarbons or HCFCs. They are called HCFCs because the molecules contain the elements hydrogen (H), chlorine (Cl), fluorine (F), and carbon (C). For the electronics industry, the two most important HCFCs at present as potential solvents are HCFC-141b and HCFC-123.

3.4.2.   The Benchmark/Phase 2 Test

To speed the approval of new and alternative technologies to replace CFC-113 and other solvents known to deplete stratospheric ozone, a group of professionals from the EPA, DOD, and industry met during 1988 to define a test procedure and a test assembly by which new technologies could be evaluated. The main driver behind these meetings (the EPA/DOD/Industry Ad Hoc Solvent Working Group meetings) was to affect a change in the military specifications regarding cleaning and defluxing technologies since the military specification are de facto standards throughout the industry, both here in the United States and also abroad.

By the end of 1988 this Ad Hoc Solvent Working Group did define a test procedure[49] and a test board so that new technologies could be evaluated. See Figure 3.17 which depicts a bare Benchmark/Phase 2 test board. The cleaning solvent used in the Benchmark (Phase 1) test was the stabilized azeotrope of CFC-113 and methanol. This test was performed independently at two different government facilities. The Electronics Manufacturing Productivity Facility (EMPF) in Ridgecrest, California and the Naval Avionics Center (NAC) in Indianapolis, Indiana. The test was performed in duplicate at both facilities. Figure 3.18 shows a soldered and populated Benchmark/Phase 2 test board. Testing at both facilities was completed by the end of February 1989.

The Phase 1 (Benchmark) test only addressed the issue of cleanliness. It did not address the issue of long-term reliability or safety and health issues. Although the Ad Hoc Solvents Working Group recognized that these latter issues were of definite importance, for the sake of getting the testing underway it deferred these issues until they could properly be addressed later. The Phase 1 test establishes, then, only a cleanliness effectiveness reference level based on defluxing with the stabilized azeotrope of CFC-113 and methanol. Results of this test were reported at the Spring 1989 IPC meeting in Orlando, Florida. These results and the test procedure can be obtained through the IPC in Lincolnwood, Illinois.[50]

The Phase 2 test is identical to the Phase 1 test except that it involves a change in the cleaning process. The types of materials that fall under the Phase 2 category are:

1. Alternative solvent defluxing—both HCFCs and solvents based on 1,1,1-trichloroethane
2. Aqueous defluxing using a saponifier to remove both the rosin flux and rosin paste required by the testing protocol
3. Semi-aqueous defluxing using a suitable material to remove the rosin flux

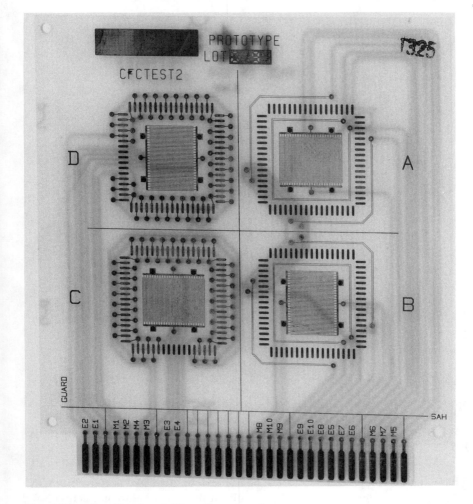

Fig. 3.17. Bare Benchmark/Phase 2 test board.

and rosin paste followed by an aqueous rinse, e.g., a defluxing agent based on a terpene

Concomitant with an alternative material, the Phase 2 test may involve a different piece of defluxing equipment. It is incumbent upon each manufacturer having a suitable defluxing process technology and who wishes to perform Phase 2 to do so on his own or to arrange for a suitable laboratory to do this.

The output for Phase 2 will be a defluxing process (cleaning agent and/or equipment) which equals or surpasses the Benchmark cleanliness reference level

Fig. 3.18.   Populated Benchmark/Phase 2 test board with leadless ceramic chip carriers (LCCCs) with no internal circuitry.

based on the test procedure, the test board, and the stabilized azeotrope of CFC-113 and methanol. The HCFC-based solvent discussed below in Section 3.4.3 successfully passed the Phase 2 test.[51] The cleaning process was identical to that employed in the Phase 1 test; the only difference was that the nitromethane-stabilized azeotrope of HCFC-141b, HCFC-123, and methanol was used in place of the nitromethane-stabilized azeotrope of CFC-113 and methanol. See Table 3.7.

Phase 3 will open the testing of such alternative technologies as:

1. Water soluble fluxes and pastes
2. Low solids "no clean" fluxes
3. Other processes such as inert blanket wave soldering involving no cleaning

Since this chapter only addresses solvent cleaning and defluxing, only alternative

Table 3.7. Phase 2 Results of an HCFC-14lb-based Solvent vs. a CFC-113-based Solvent in a Two-Sump Batch Degreaser.

IONIC CONTAMINATION RESIDUE
(μg NaCl/in^2)

	PHASE 1 STABILIZED CFC-113/METHANOL AZEOTROPE		PHASE 2 STABILIZED HCFC-14lb/HCFC-123/ METHANOL/AZEOTROPE*	
PROCESSES SEQUENCE	ARITHMETIC MEAN	UPPER LEVEL FOR VARIANCE	ARITHMETIC MEAN	SAMPLE-TO-SAMPLE VARIANCE
A	2.0	1.7	0.6	1.0
C	3.8	6.7	2.5	0.1
D	10.7	10.1	8.1	5.7

ROSIN CONTAMINATION RESIDUE (μg)

	PHASE 1 STABILIZED CFC-113/METHANOL AZEOTROPE		PHASE 2 STABILIZED/HCFC-14lb/HCFC-123/ METHANOL AZEOTROPE*	
PROCESS SEQUENCE	ARITHMETIC MEAN	UPPER LEVEL FOR VARIANCEE	ARITHMETIC MEAN	SAMPLE-TO-SAMPLE VARIANCE
A	301	44,135	47	183
C	3,135	535,961	3,128	72,578
D	3,945	13,696,891	1,536	230,767

*Composition in weight percent: HCFC-14lb 86.1; HCFC-123 10.0; methanol 3.6; nitromethane 0.3. For further details, see References 50 and 51.

solvent defluxing will be addressed here. Other chapters in this book will take up other aspects of Phase 2 and Phase 3 testing.

3.4.3.   Alternative Defluxing Solvents

To replace defluxing solvents based on CFC-113, preliminary tests indicate that HCFC-141b-based solvents are good defluxing materials. It was discovered also that the addition of a small amount of HCFC-123 suppresses the flame limits of HCFC-141b-based solvents. Hence, the most promising solvent of the HCFC-141b family of solvents is one containing both HCFC-141b and HCFC-123 with the addition of methanol. The ODP of this solvent is 0.07 to 0.06 depending on the amount of HCFC-123 in the formulation. The physical properties of HCFC-141b and HCFC-123 are given in Table 3.8. The properties of the stabilized solvent based on these two ingredients plus methanol will be supplied by the manufacturers. The structural formulas of HCFC-141b and HCFC-123 are depicted in Figure 3.19.

Table 3.8. Physical Properties of HCFCs Used in Defluxing.

	HCFC-141b	HCFC-123
Molecular weight	116.9	152.9
Appearance	Colorless liquid	Colorless liquid
Boiling point,		
°F	89.7	82.2
°C	32.0	27.9
Vapor pressure @ 77° F (25° C), Psia	11.6	13.2
Liquid density @ 77° F (25° C),		
g/cc	1.236	1.463
lbs/gal	10.28	12.20
Latent heat of vaporization at NBP, Btu/lb	74.0	95.4
Solubility of water @ 77°F (25°C), ppm	420	660
Solubility in water @ 77°F (25°C), ppm	660	2100
Liquid viscosity @ 77°F (25°C), cP	0.43	0.48
Vapor flammability @ 77°F (25°C)		
Lower limit, vol.%	7.6	None
Upper limit, vol.%	17.7	None
Flashpoint	None	None

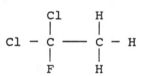

(1)   1,1-dichloro-1-fluoroethane (HCFC - 141b)

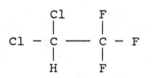

(2)   1,1-dichloro-2,2,2-trifluoroethane (HCFC - 123)

Fig. 3.19.   Structures of (1) 1,1-dichloro-1-fluoroethane (HCFC-141b), (2) 1,1-dichloro-2,2,2-trifluoroethane (HCFC-123).

Table 3.9. Defluxing Results of an HCFC-14lb-based Solvent versus a CFC-113-Based Solvent in a Two-Sump Batch Degreaser.

IONIC CONTAMINATION RESIDUE, μg NaCl/in.^{2}

	(PWAs) FLUX					(SMAs PASTE)	
SOLVENT	#1	#2	#3	#4	#5	#1	#2
Stabilized HCFC-14lb solvent with HCFC-123 and methanol	6.7	6.3	9.7	12.9	9.9	6.7	11.0
Stabilized CFC-113 solvent containing methanol and other ingredients	9.1	7.0	13.9	9.9	14.2	12.3	9.6

ORGANIC CONTAMINATION RESIDUE, μg ROSIN/ in^2

	(PWAs) FLUX				(SMAs PASTE)	
SOLVENT	#1	#2	#3	#4	#1	#2
Stabilized HCFC-14lb Solvent with HCFC-123 and methanol	139	109	130	147	283	38
Stabilized CFC-113 Solvent containing methanol and other ingredients	123	162	138	154	404	88

Note: Flux #1 is RA-MIL; Fluxes #2–#4 are RA; Flux #5 is SA. Pastes #1–#2 are RA.

Defluxing results based on a solvent composed of HCFC-141b, HCFC-123 and methanol have been conducted. This material shows equal if not improved ability when compared to CFC-113-based solvents for removing both flux and paste residues.[52–55] Performance evaluations of this solvent were conducted in both a batch defluxer and a conveyorized in-line defluxer. For example, the results (averages for four boards per run) for defluxing conventional PWAs and SMAs in a batch defluxer are presented in Table 3.9. See Figure 3.20 which depicts the conventional PWA used in this experiment; Figure 3.21 shows the SMA used.

The solvent based on HCFC-141b was the stabilized azeotrope containing HCFC-141b, HCFC-123, and methanol. The other solvent was a stabilized azeotrope of CFC-113, methanol, and several other ingredients. The details of the cleaning cycles are given in Reference 55. The results are (1) ionic results using a conventional ionic contamination tester and are expressed in micrograms of sodium chloride or equivalent per square inch (μg NaCl/in.^{2}) and (2) organic (rosin residue) results using a UV/VIS-spectrophotometer and are given in micrograms of rosin or equivalent per square inch (μg rosin/in.^{2}). Flux #5, the SA flux, did not absorb in the UV/VIS region; hence, it does not appear under the results for residual rosin. It can be seen that the solvent based on HCFC-

Fig. 3.20.   Conventional printed wiring assembly (PWA) with throughhole components (THCs).

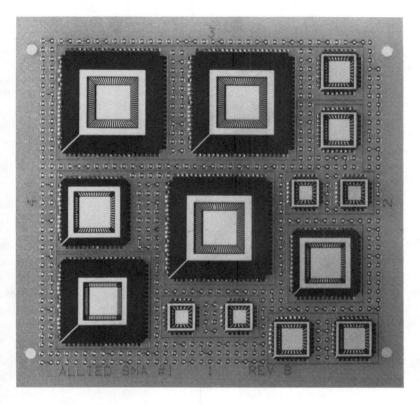

Fig. 3.21.   Surface mount assembly (SMA) with leadless ceramic chip carriers (LCCCs) with no internal circuitry.

141b performs as well, if not better, than the conventional CFC-113-based solvent.

In Table 3.10 results (averages of four boards per run) are given for cleaning PWAs and SMAs run in a conveyorized, inline defluxer at 3 feet per minute. The same two solvents as described in the paragraph above were used. Three different fluxes and two different pastes were employed. Again, the details of the cleaning cycles are given in Reference 55. The results are expressed in micrograms of sodium chloride or equivalent per square inch (μg NaCl/in.^{2}). The PWAs were tested in the conventional ionic contamination testers; the SMAs in the new type of ionic contamination tester containing spray headers in the test tank and heated test solution (see Figure 3.9). Again it can be seen that the solvent based on HCFC-141b performs as well if not better than the conventional CFC-113-based solvent.

For using the HCFC-141b/HCFC-123/methanol azeotrope in an inline de-

Table 3.10. Defluxing Results of an HCFC-141b-based Solvent vs. a CFC-113-based Solvent in an Inline Liquid Seal Defluxer (3 ft/min.)

| | IONIC CONTAMINATION RESIDUE | | | | |
| | (PWAs) FLUX | | | (SMAs PASTE) | |
SOLVENT	#1	#2	#3	#1	#2
Stabilized HCFC-141b Solvent with HCFC-123 and methanol	4.1	8.2	9.7	6.1	5.8
Stabilized CFC-113 Solvent containing methanol and other ingredients	5.4	8.3	7.9	8.9	5.9

Note: Flux #1 is RMA; Fluxes #2–#3 are RA. Paste #1 is RA; Paste #3 is RMa.

fluxer, it would be best to incorporate increased vapor retention designs since the solvent boils at 85°F. If this is not feasible, the machine should be equipped with finned tubing for both its primary condensing coil and entrance and exit end vapor traps. A batch cleaner should have a minimum of 100% freeboard to contain the new solvent. Suitable low emissions equipment, both batch and inline, is available to adequately contain the new solvents based on HCFC-141b.

Safety Aspects of HCFC-based Solvents.   Any new molecule must be subjected to an extensive battery of toxicological tests. As of 15 December 1988, the following toxicology tests using HCFC-141b have been completed: acute inhalation study, 2-week subacute inhalation study, and rat teratology study [negative]. An extensive 13-week subchronic inhalation study is being conducted and should be completed by March. At present, the permissible exposure limit (PEL) for HCFC-141b has been set at 500 ppm. The PEL is, of course, an interim or suggested exposure level. After extensive, long-term testing (two year toxicological test) is completed, a threshold limit value (TLV) will be assigned. The HCFC-141/b/methanol/nitromethane azeotrope does not exhibit any flash-point according to the Tag Open-Cup Test (ASTM D1310-86).

3.5.   CONCLUSION

Solvent defluxing has been used for years in the electronics industry for successfully cleaning conventional throughhole PWAs and the newer SMAs being produced today. The technology is well understood by electronics manufacturing engineers, and the results that can be achieved are good to excellent provided the right solvent, equipment, and cleaning cycle are chosen. Because a wide experience base exists for solvent defluxing technology, a high degree of confidence can be placed on its continued usage to satisfy the stringent demands of the electronics industry, particularly for defluxing surface mount assemblies. With the advent of fine pitch technology (FPT), the demand for adequate defluxing will even be greater. With the introduction of a new generation of solvents based on HCFC molecules, solvent defluxing should prove to be a viable cleaning option for the future while still meeting the demands for an environmentally acceptable material.

REFERENCES

1. Prasad, R. P. *Surface Mount Technology: Principles and Practice*. New York: Van Nostrand Reinhold, 1989, pp. 7–11.
2. Manko, H. H. *Solders and Soldering*, 2nd ed. New York: McGraw-Hill, 1979, pp. 1–3.
3. Bonner, J. K. "Flux Fingerprinted by Ionic Content," *Cir. Mfg.*, 57–58 (Jan 1987).
4. Enos, H. J., Jr., G. C. Harris, and G. W. Hedrick, *Kirk-Othmer Encyclopedia of Chemical Technology*, 2nd ed., vol. 17, "Rosin and Rosin Derivatives." New York: John Wiley and Sons, 1968, pp. 475–508.
5. Genge, G. A. "Resin Acids: Analysis by Mass Spectrometer (as Methyl Esters)," *Anal. Chem.*, **31**(11) 1750 (Nov. 1959).
6. Manko, H. H. Op. cit., p. 14.
7. Bonner, J. K. "A Few Words about Solvents," IPC Technical Paper 302, Sept. 1979.
8. Tautscher, C. J. *The Contamination of Printed Wiring Boards and Assemblies* Bothell, WA: Omega Scientific Services, 1976, Chap. 5, pp. 6–22.
9. Abbott, W. H. "Corrosion Still Plagues Electronic Packaging," *Elec. Pack. & Prod.*, 28–33 (Aug. 1989).
10. Tautscher, C. J. Op. cit., Chap. 6, 4–15.
11. Lovering, D. G. "Rosin Acids React to Form Tan Residues," *Elec. Pack. & Prod.*, 232–234 (Feb. 1985).
12. Tatone, M. J., and K. J. Bradford. "White Residue on Printed Wiring Board Assemblies—A Problem for the Electronic Industry?," *Soc. Mfg. Engrs.*, EE76-583 (1976).
13. Bonner, J. K. "White Residue—Questions and Answers," unpublished paper, 1989.
14. Westerlaken, E. "Rosin Solder Flux Residues Shape Solvent Cleaning Requirements," *Elec. Pack. & Prod.*, 118–124 (Feb. 1985).
15. Bonner, J. K. "A Few Words about Solvents," Op. cit. 8–11.
16. Ibid., p. 2.
17. Sanger, D., and K. Johnson. "A Study of Solvent and Aqueous Cleaning of Fluxes," Naval Weapons Center Technical Paper 6427, 96–98, Feb. 1983.
18. Anon. "Toxicological and Carcinogenic Evaluation of a 1,1,1-Trichloroethane Formulation by Chronic Inhalation in Rats," Dow Research Laboratory Toxicology Report, Midland, MI, Oct. 1978.
19. Heath, B. A. "Inline High Pressure Cleaning of Fine Pitch Surface Mount Assemblies," *Proc. Nepcon West '88*, 496–509, Feb. 1988.
20. Egan, T. F. "Determination of Plating Salt Residues," *Plating*, 350–354, April 1973.
21. DeNoon, R. J. and W. T. Hobson. "Printed Wiring Assemblies; Detection of Ionic Contaminants on," Naval Avionics Facility Materials Research Report No. 3-72, 1972.
22. Bonner, J. K. "A Comparison of Direct Surface Analysis Techniques with Solvent Extraction/Contaminant Profiling Techniques for Ascertaining Production Cleanliness of Printed Wiring," IPC Technical Paper 323, 5–21, April 1980.
23. Knopeck, G. M., and J. K. Bonner, "A Critique of Ionic Contamination Testing for Surface Mount Technology," *Proc. Nepcon West '88*, 519–528, Feb. 1988.

24. Jupille, T., D. Burge, and D. Togami, "Ion Chromatography Uses Only One Column to Get All the Ions," *Res. & Dev.*, March 1984.
25. Wargotz, W. B. "Ion Chromatography Quantification of Contaminant Ions in Water Extracts of Printed Wiring," IPC Technical Paper 248, Sept. 1978.
26. Bonner, J. K. "A Comparison of Direct Surface Analysis Techniques with Solvent Extraction/Contaminant Profiling Techniques," 27–32.
27. Bonner, J. K. "An Analysis of Four Printed Wiring Assemblies Using the Contaminant Profiling System," IPC Technical Paper 644, n.d.
28. Wagner, L. K. "Organic Surface Contamination—Its Identification, Characterization, Removal, Effects on Conformal Coating," Boeing Military Airplane Co. Report, Oct. 1981.
29. Kaier, R. J. "A Simple Method for the Detection and Identification of Organic Non-Ionic Contaminants on PWB and PWA Surfaces," IPC Technical Paper 468, April 1983.
30. Klima, R. F., and J. K. Bonner. "The Evaluation of PWA and SMA Cleanliness Levels for 'In-line' Defluxing by High Performance Liquid Chromatography," *Proc. Nepcon West '89*, Vol. 2, 1370–1389, March 1989.
31. Klima, R. F., and J. K. Bonner. "White Residue Analysis by High Performance Liquid Chromatography," IPC Technical Paper 713, April 1988.
32. Richey, W. F., J. A. Trombka, E. L. Tasset, T. D. Cabelka, and A. H. Hazlitt. "New Analyses for Residual Rosin on Cleaned Electronic Circuit Boards," *Proc. Nepcon West '85*, 301–313, Feb. 1985.
33. Bonner, J. K. and J. M. Lewis, "Coulometry: A Promising Method for Quantifying Organic Residue on SMCs after Cleaning," *Proc. Tech. Conf.*, 7th Ann. I.E.P.S. Conf., 569–580, Nov. 1987.
34. Richey, W. F., et al. "New Analyses for Residual Rosin."
35. Bonner, J. K., "A Comparison of Four Solvents for Defluxing Printed Wiring Assemblies," *Proc. Circuit Expo '86*, 94–103, Oct. 1986.
36. Gorondy, E. J. "Surface Insulation Resistance—Part I: The Development of an Automated SIR Measurement Technique," IPC Technical Paper 518, Sept. 1984.
37. Gorondy, E. J. "Surface Insulation Resistance [SIR] Part II: Exploring the Correlation between Standard Industry and Military 'Sir' Test Patterns—A Status Report," IPC Technical Paper 543, Apr. 1985.
38. Gorondy, E. J. "Surface/Moisture Insulation Resistance (SIR/MIR)—PART III: Analysis of the Effects of the Test Parameters and Environmental Conditions on Test Results," IPC Technical Paper, Oct. 1988.
39. Zado, F. M. "Electrical/Electronic Reliability Considerations in Modern PWB Manufacturing and Assembly Operations," Printed Circuit World Convention III Technical Paper 70, May 1984.
40. Willson, D. L. "The Trials and Tribulations of Implementing Surface Insulation Resistance Testing," *Proc. Electronics Mfg. Sem.*, 12th Ann. EMPF/Naval Weapons Center, China Lake, CA Technical Seminar, 137–159, Feb. 1988.
41. Heffner, K. H., and J. C. Brand; B. Grosso and D. Terrell. "The Effect of Flux Residues on Long and Short Term PWB Insulation Resistance Assays," IPC Technical Paper 680, Apr. 1988.
42. Osborne, A. G. "Modern Thoughts on Surface Resistance," IPC Technical Paper 498, Sept. 1983.
43. Jennings, C. W. "Insulation Resistance Measurements of Coated and Uncoated Printed Wiring Boards," IPC Technical Paper 496, Sept. 1983.
44. Der Marderosian, A. "Raw Material Evaluation through Moisture Resistance Testing," IPC Technical Paper 125, Sept. 1976.
45. Monastersky, R. "Antarctic Ozone Reaches Lowest Levels," *Sci. News*, **132**(10/10/87), 230.

46. Monastersky, R., "Arctic Ozone: Signs of Chemical Destruction," *Sci. News,* **133**(11/11/88), 383.
47. Monastersky, R. "Clouds without a Silver Lining," *Sci. News,* **134**(10/15/88), 249–251.
48. Anon. *Montreal Protocol on Substances That Deplete the Ozone Layer (Final Act),* 1987.
49. Anon. "Cleaning and Cleanliness Testing Program: A Joint Industry/Military/EPA Program to Evaluate Alternatives to Chlorofluorocarbons (CFCs) for Printed Board Assembly Cleaning," IPC, March 30, 1989.
50. Anon. "Cleaning and Cleanliness Test Program: Phase 1 Test Results," IPC Technical Report 580, Oct. 1989.
51. Bonner, J. K. "Phase 2 Test: Final Report/Allied-Signal Inc.," Allied-Signal Report, Jan. 1990.
52. Bonner, J. K. "Cleaning Surface Mount Assemblies: The Challenge of Finding a Substitute for CFC-113," *Proc. Nepcon West '89,* Vol. 2, 1051–1058, March 1989.
53. Merchant, A. N., and M. C. Wolff, "Screening and Development of Ozone/Greenhouse Compatible Cleaning Agents," *Proc. Nepcon West '89,* Vol. 1, 756–758, March 1989.
54. Bonner, J. K. "New Solvent Alternatives," *Printed Circ. Ass.,* **3**(9), 36–38 (Sept. 1989.
55. Bonner, J. K. "Solvent Alternatives for Electronics for the 1990s." *Proc. Nepcon West '90,* Vol 2, 1601–1608, Feb. 1990.

4
Aqueous Defluxing:
Materials, Processes, and Equipment

Gregory C. Munie
AT&T Bell Laboratories

4.1. INTRODUCTION

4.1.1. Why Aqueous?

It is easy to predict the future. To be accurate in one's predictions is another story. At the risk of sounding either too smug or too prophetic I believe I can say that aqueous cleaning for electronics is an idea whose time has come. The industry as it is configured now with a heavy dependence on solvent based processes will change. The driving force for these changes will be external, regulatory pressure, and internal, crises of conscience: there is weighty evidence that the way we live now may ultimately result in our demise as a species. As there are few volunteers for a change in modern lifestyle it therefore behooves us to change the ways to maintain that lifestyle. Aqueous processing is one of the attractive (and relatively painless) alternatives.[1]

This chapter will discuss the issues associated with aqueous processing: materials, process, and equipment. As such it will draw heavily from my experience with aqueous cleaning and from the published literature on the subject. It is not intended to be a complete treatise on aqueous cleaning but to give the reader an idea of the possibilities and pitfalls associated with such processing.

---

[1]At present the effect on the environment of a particular choice is not always well defined. For example, while aqueous processing eliminates known concerns about ozone depletion and most worries about low level air pollution it may have a negative impact (in some cases) on global warming and water quality. My assessment of the present state of aqueous cleaning is a positive one. It is not the final one.

4.1.2. Plusses and Minuses

No process is perfect. Aqueous processing is no exception. To begin the task of describing the basics of aqueous cleaning it is worthwhile to look at the good and the bad, the easy and difficult of this technology.

On the positive side aqueous processing offers the possibilities of environmental compliance, low cost operation, and assurance of reliability.

Water is an environmentally acceptable material. The technologies for processing water are well developed and, for most types of pollution problems, do not demand (in my opinion) discovery and development so much as adaptation. It is safe to say that, technically and politically, water based technology is acceptable from an environmental standpoint.

On the cost side, water is still relatively cheap, even counting costs of treatment. Some studies of the relative costs have shown that aqueous cleaning is equal or less in cost than solvent based systems (Goldstein 1980, Ellis 1986, Manko 1986). The consensus report of a panel convened for discussion of aqueous cleaning for military electronics (Manko and Ayers 1989) was that aqueous cleaning is comparable to the cost of solvent systems.

To be fair, some articles have cited higher costs (Osterman 1986, Kenyon 1978). But at this point in time, with increased restrictions on emission and on disposal of spent solvent materials (including landfill bans), solvent based processes can be expected to quickly outstrip aqueous processes in cost. This cost of process does not include any capital investment that a manufacturer need make in order to achieve a solvent to aqueous conversion.

Much of the discussion as to cost comparison has centered on the heating and drying processes used in an aqueous system. A recent study (Askengren 1989) has claimed that the high cost of energy associated with aqueous processing will result in process costs 20–25% higher for water systems than for solvent systems. This study was for metal degreasing operations. Studies within AT&T for defluxing have shown the reverse (Munie, Owens, Sharp, and Wenger 1990). At the present stage I believe that costs are essentially equivalent (ICF 1987, UNEP 1989). A complete breakdown of costs due to all factors will probably never be forthcoming due to the wide range of process conditions and treatment options. My advice to those readers contemplating a change to aqueous would be to examine for themselves the comparative costs for *their specific process*. I would predict (an author's prerogative) the costs will differ either for or against the aqueous process by no more than 10%. If the difference is greater, it would be worth the effort to determine the exact contributor to that difference and share it with others in the industry to advance the resolution of the debate.

The question of reliability has long been debated in electronics. The general feeling is that the presence of ionic material residues constitute the greatest reliability problem. The use of water as a final rinse to insure reliability in solvent based cleaning systems (Sanger and Johnson 1983) is an accepted military cleaning practice. An all aqueous process can be expected to give more than adequate cleaning of ionic residues and thus eliminate one of the main reliability hazards.[2]

For every positive there is a negative. (A Yin for a Yang if you would.) On the negative side of aqueous processing one finds difficulties in the handling of spent aqueous effluent, design compatibility, adequate drying of assemblies, and complete removal of nonionic materials.

The move currently in legislation is to address the use and emission of solvents. But those now looking at aqueous materials would be wise to see this as only part of a larger trend in environmental legislative activity. Water quality is definitely under scrutiny. In putting in an aqueous system it is vital that the local authorities be consulted, that adequate provision be made for treatment and disposal of the effluent (Sheldahl and Horton 1987), and that the possibility of additional restrictions on processing be kept in mind. An assembler may want to consider the installation of a closed loop system (Goldstein 1980, Munie, et. al., 1990).

One advantage (and environmental disadvantage) of solvent systems is their volatility. Water, being relatively nonvolatile, requires provision for removal from the assembly processed. Although most commercial aqueous cleaners usually have adequate drying, certain types of assemblies (notably surface mount) may require special care in drying to prevent test difficulties. (A discussion on the issues of assembly drying is presented later on in this chapter.)

A transition from solvent to aqueous may not always be immediately possible. Conversion of process may involve only machine replacement. But designs that are in production may require redesign to insure that they are compatible with the new process. A more complete discussion of this issue will appear in the next section.

Whereas water is a superior remover of ionic materials, solvents are superior in removal of nonionics. A hydrophobic,[3] nonionic residue such as rosin may not be a corrosive reliability hazard. It may however interfere with testing of the assembly, increase the resistance of connector contacts, or interfere with proper adhesion of conformal coatings (Tautscher 1976). If removal of hydrophobic nonionic material is required, water must be supplemented with additional

---

[2]This is not to say that aqueous processing is hazard free. The reliability and performance problems associated with water cleaning will be discussed shortly. So while ionic materials are among the most talked about (and tested for) contaminants, nonionics can also have an impact on circuit performance. This may be through conductivity (Kenyon 1979b) or through effects on conformal coating (Tautscher 1976, Ellis 1986)

[3]The term *hydrophobic* is used here to distinguish the residue from the nonionic residues left by surfactants. These latter residues are water soluble (hydrophilic), whereas a hydrophobic residue such as rosin is not.

Table 4.1. Examples of Typical Design Considerations.

CHOICE	CONSIDERATIONS
Functionality	Reliability, including choice/impact of soldering, and cleaning processes
Components	Process compatibility/choice, reliability, and cost
Flux choice	Compatibility with components, soldering, and cleaning process
Solder process	Compatibility with components and cleaning process
Cleaning process	Compatibility with components and flux process; disposal of effluent; effect on test and coatings
Test Process	Compatibility with cleaning
Coating Process	Compatibility with flux and cleaning

cleaning agents. This issue will be dealt with in the areas below on process choice and testing of assemblies.

As a final note to the above plusses and minuses, remember: every process has its good and bad. How an assembler deals with the tradeoffs between the two is in theory the same for both solvent and aqueous processes. In practice the process window of operations as defined by criteria such as defect rate and reliability will be different for different processes. Thus when an assembler switches from a solvent to an aqueous process no one should be surprised if the change involves more than the substitution of one cleaning machine for another.

4.2.   DESIGN CONSIDERATIONS

Design is usually considered to cover the nuts and bolts of an assembly: the what goes where to do what. For the purpose of adequately covering the topic of aqueous cleaning I must extend the concept of design to include the process. For a manufacturer to be really efficient and have a chance at being successful in the competitive world, an integrating of the traditional design concept and the assembly process is vital. And the assembly process should not be considered to stop at the shipping dock. As was pointed out previously, a plus of aqueous cleaning is its environmental compatibility. Design should thus encompass what the product looks like, what it does, how it is to be manufactured, and what is to be done with the "leftovers" of the assembly process. Table 4.1 lists some of the considerations involved in the overall design. Note how all the sections listed are interrelated.

As an exercise the reader should consider his or her own process, using the table as a (crude) guide. How does each step in your process impact the up and downstream steps? Consider how the change of one step from flux A to flux B or cleaning process X to cleaning process Y would impact the overall considerations of reliability or environmental impact. An assembler (or team of marketing, design, process, and environmental engineers) who sets these requirements at the beginning of the design step and integrates them for total optimization holds the key to success.

To discuss design, the topic will be divided into four separate topics: layout, components, surface mount versus throughhole, and process/material considerations. For additional detailed treatment of some design and assembly process considerations see Chapter 1 of this book.

4.2.1. Layout

The layout of an assembly impacts the success of an aqueous process. The two key points in this part of the design process are irrigability and dryability/rinsability.

Irrigability has been long discussed. The distance a part stands off the board in an assembly affects the ability of the cleaning media to clean under that part. The debate over why one medium can or cannot adequately clean under a part has often centered on surface tension. Attempts have also been made to derive various numerical values for the interactions of cleaning materials with residues and with standoff distances (Kenyon 1979a, Brous and Schneider 1984, Comerford 1984, Cabelka, Archer, and Trombka 1986, Wang 1987, Trieber 1989). It is the author's experience that any of these derived formulas or evaluations are at best rough guidelines. For anyone setting up a cleaning system of any type, the considerations of equipment, processing conditions and time, residue type, cleaning media, and standoff/irrigability all play a part in whether a process will provide the level of cleaning required. In general, however, the farther a part stands off the circuit board and the fewer mechanical obstructions lie between the dirt and the cleaning medium the more chance any cleaning media in any process will have of removing that dirt. The basic rules are thus:

1. In standoff, more is better.
2. The larger the area a device covers the more standoff is required for adequate cleaning. For example, a small ceramic surface mount chip capacitor may need no more than the standoff given it by its solder connections. A large leadless chip carrier may be almost impossible to completely clean under.
3. Configurations that trap flux reside should be avoided. An example of this is a pattern of adhesive dots for surface mount devices that results in a flux "cul-de-sac."
4. Closely spaced leads on a part or closely spaced parts on a board can impede the flow of cleaning media. (It has been my experience that these effects are small relative to the ones of standoff and "blind alleys" in design.)

5. Large standoffs do not automatically guarantee cleaning. As will be discussed blow, cleaning media must be matched to the flux used.
6. The cleaning equipment must be adequate to the task.
7. And finally know what and why you are cleaning:

A. Match the cleaning medium to the dirt.
B. Determine beforehand what level of cleanliness the assembly needs. Consider whether that level is set by:

a. Customer's specifications.
b. Field reliability requirements. (In this case the presence of some types of residues, e.g. pure rosin, may be extraneous to assembly performance.)
c. The next step in the process, e.g. test or conformal coating.

After the cleaning part of the aqueous process is complete, the assembly must receive a final rinse and be dried. To a certain extent these requirements overlap somewhat with standoff. Whereas standoff deals with free flow of cleaning media, the questions of rinsability/dryability deal with free flow of the final rinse water and the free flow of a drying gas (usually air) in and residual water out. (This gives the layout designer a second reason to avoid cul-de-sacs.) Note that one common example of an entrapping part is the surface mount ceramic chip. Just as this part often traps flux, so too can it entrap water. The removal of the flux may require thought as to choice of flux, cleaning media, and equipment. The removal of the water may simply be a matter of equipment choice and process conditions, e.g. time at temperature that the assembly can tolerate.

4.2.2. Component Choice

If you are now using a solvent process, you may or may not be able to make a direct transition in the assembly process from solvent to aqueous. The compatibility of the components with an aqueous process must be considered. That in turn can be broken down into questions of reliability and entrapment.

From the reliability standpoint components must be able to withstand exposure to the cleaning medium. (Note that this assumes no change in flux.) There should be no deleterious effects either short or long term.

Parts should not entrap final rinse water or cleaning medium. This has been dealt with above in reference to board layout and standoff of the parts. Entrapped fluid, either in the body of the component or in a nondraining part, e.g., some types of connectors, is not a desirable situation.

4.2.3.   Surface Mount versus Throughhole

Surface mount technology usually increases the difficulties of cleaning. A listing of the reasons for this increase in difficulty includes:

1. Generally lower standoff of surface mount devices.
2. Increased density of parts on the assembly.
3. And, in some cases, longer solder process times that result in more intractable residues.

4.2.4.   Process/Materials Considerations

As was pointed out earlier, water as a cleaning medium is ideal for ionics, yet somewhat at a disadvantage as a remover of nonionics. A chemist usually describes this as "like dissolves like." The same is true in processing. The various sequential operations should be matched with each other for optimum yield. Applying this to the cleaning operation means matching dirt with the cleaning medium and process. Studies have shown that for the same cleaning medium results can vary greatly depending on the flux used (Sanger and Johnson 1983). Good references on the matching of dirt to cleaning can be found in the IPC cleaning manual, IPC-C-40 and Ellis, 1986 and in Chapter 1 of this text.

4.2.5.   Summary

The above discussion has been written not to give all the specifics associated with design of assemblies but to give the reader an idea of some of the pitfalls in the interaction of design and aqueous processing. However, it must be stressed again: design encompasses more than what the final assembly looks like or does. The considerations involved in carrying out a successful manufacturing process, and in particular for successful cleaning using aqueous processing, must involve more than just the appearance or functionality of a part. The "how" a part is made should mesh with the "why."

4.3.   PROCESS CONSIDERATIONS

In this section some specifics of cleaning materials and equipment will be discussed. Of necessity some of this material will overlap with other chapters in this volume. At appropriate times the reader's attention will be directed to those cross references.

To begin, I will review the usual flow of a cleaning process (I say usual because there may be exceptions to this flow) and briefly discuss the process parameters that influence cleanliness.

Starting with something that requires cleaning (in our case an electronic assembly) the process flow involves, in order:

1. A wash cycle. This may involve either pure water or some combination of water and a cleaning agent that interacts in some way with the "dirt" to be removed. This interaction may be a chemical reaction in which a new compound consisting of the reaction products of the dirt and cleaning agent is formed. It could also be a physical or a physical-chemical type interaction. An example of the former would be the physical blasting away of particulate matter by the water. (Menon et al. 1988) An example of the latter would be the solvation of ionic contamination in water.
2. A rinse cycle. The reaction products, the solvated dirt, and/or the suspended particulate matter, is replaced with pure water, i.e., without any other material in an amount that might affect the final assembly in an undesired manner.
3. A dry cycle. The water is removed from the assembly. Examples of removal methods are evaporation of the water and physical removal by jets of air or some other gas.

It is at the completion of all these steps that the assembler (hopefully) is blessed with a clean, reliable assembly ready for the next step.

In regard to the process parameters that most directly affect cleanliness it has been my experience that there are really only four which impact on cleanliness.[4] These are time, temperature, medium, and flow/volume used.

If you have an infinite amount of time you can probably achieve (assuming no gross process mismatches) any level of cleanliness you desire. Fortunately or unfortunately none of us is so blessed or cursed. The process engineer must balance the level of cleanliness desired and time in process. In most cases where the other factors influencing cleaning are not at one extreme or the other and where cleanliness down to semiconductor levels is not required, the requirements for balance are within reason for profitable operation of the company. If however, an engineer is cursed with an intractable flux/solder process, an extremely dense design, and the necessity of reaching a very high level of cleanliness, the time to reach the desired cleanliness level may become a problem for smooth process operation.

Increasing the process temperature decreases the cleaning time required and increases the achievable level of cleanliness given a fixed process time by in-

---

[4]This in no way implies that cleanliness is boiled down to control of only four items. How we might wish for a process with only four dials to be set to attain perfect results! What I describe here are the parameters that—all other factors being accounted for, e.g., board design, flux type, components, etc.,—are adjustable in such a way as to optimize the cleaning process.

creasing reaction rates and solubility. That this process is not open ended for saponification reactions has been demonstrated by Ellis (1986). The behavior of any surfactants used in the process is also a function of temperature. In this case the detergency of a surfactant system has a maximum near the transition temperature referred to as the surfactants cloud point (Rosen 1978). In general however, increases in temperature (up to a point) will tend to decrease the time required for cleaning.

The cleaning medium directly affects the ability to clean. The first concern is to match the medium to the dirt. As mentioned earlier (see Section 4.2.4.), guidelines exist for such matches. If, for example, a rosin flux is used, then a saponifier is appropriate in the wash section of the cleaning process. The level of the saponifier concentration (e.g., Ellis 1986) and the rate of changeover of the saponifier bath then become the governing parameters in control of the media. For any process, however, the ultimate variable, as far as the final performance of the assembly is concerned, is water quality, e.g., IPC-AC-62. Depending on the final level of cleanliness required, the water an engineer can use may be either tap water, softened water, or deionized water. Tap water may be sufficient but should not be expected to give the same results as deionized water.

The last parameter, flow/volume, is most often fixed by cleaning equipment choice. I mention it here because it has been my experience that the method of delivery of the cleaning medium to the substrate is tied in with the actual process of removal as much as temperature and time. (The reader will, I hope, forgive me for putting it here rather than under equipment considerations. My background as a physical chemist drives me to place it with the other parameters governing the cleaning "reaction.") Much has been said by equipment suppliers on the importance of pressure and/or volume in material delivery. From a particulate removal standpoint (Menon et al. 1988), pressure is, I feel, the most important. From a chemist's standpoint, where the cleaning can be viewed as a class of reaction or solvation, I feel that volume is more important. (I would argue that the presence of excess reactant/solvent for the dirt to interact with will thermodynamically drive the process toward removal.) Recent studies indicate that the two, pressure and volume, may actually be inextricably related (Bahr and Andrus 1989.)

Reader, please note: these three parameters may not be independent variables! Hot, high purity water delivered at a high flow rate to the assembly will, generally, give a higher cleanliness level than the same flow of low temperature water of lower purity. But, depending on the assembly and its condition when it reaches the cleaner, these parameters will have a particular range of values that will give optimum results.

To end the discussion at this stage would, assuming all the right choices in design and process parameter settings have been made, result in a clean but somewhat soggy assembly. Drying, the final step, must not be neglected! Drying can affect both postcleaning handling of the assemblies and, depending on the limits of process water quality, the final level of board cleanliness. Inefficient drying produces wet assemblies that may not be amenable to test after cleaning. An efficient drying procedure that removes water through nonevaporative means, e.g., air blowoff, may allow an assembler to use tap water rather than deionized water for the final rinse. A discussion of drying in aqueous processing can be found in Woodgate et al. (1980) and Elliot (1990).

4.3.1. Cleaning Materials

In aqueous processing cleaning materials can be divided into pure water, water with added cleaning agents such as neutralizers or saponifiers, and materials that involve the use of semi-aqueous processing.

Water. If you must have an aqueous process might I recommend that you try water first. Sound a little silly? Perhaps, but it best illustrates the point made previously on choice of process. Also, from a materials standpoint, it is the least expensive of the aqueous processes.

If you wish to take advantage of the best attributes of water, the process of choice is water soluble flux with water-only cleaning. The attributes of this process for both throughhole and surface mount have been described in detail (Wargotz, Guth, and Stroud 1987, Danford and Gallagher 1987, Comerford, 1984, Aspandiar, Piyarali and Prasad 1986). Adequate cleaning has been demonstrated in all these processes. Here, as noted previously, the design considerations must take into account:

1. Flux choice as determined by soldering performance, cleanability of flux residue after soldering (not of the raw flux!), and reliability of the final assembly. (Note that these three points are interrelated. A more complete discussion of this will be found in Chapter 2 of this book.)
2. Board layout for adequate irrigability. (See, for example, Comerford 1984.)
3. Component choice, which includes resistance of the package to intrusion by either the flux or the water.
4. The cleaning process itself. This will include equipment selection and settings of the parameters mentioned above.
5. Disposal of the spent effluent.

The summation of these considerations will provide one with the basis for setting up an all water process.

To a great extent these factors are interrelated. Flux choice impacts soldering, cleaning, reliability, and effluent disposal (Woodgate 1983). Board layout and

Table 4.2. Typical Materials in an Organic Saponifier.

COMPOUND CLASS	PURPOSE
Organic amines	Saponification reaction with abietic acid
Nonionic surfactants	Wetting and rinsing aids
Cellosolves and carbitols	Enhanced rosin solvation

component choice influence cleanability. In turn, cleanability may be then mitigated by choice of flux, e.g., a milk flux leaving a relatively benign residue.

It may seem that these considerations are burdensome. However, their consideration and control will insure a smoothly running process capable of delivering reliable assemblies.

Neutralizers and Saponifiers.   The use of neutralizers or saponifiers may be required if the flux is acidic or if a rosin based material that requires removal after soldering is used.

Neutralizers.   The chemistry involved in this process is primarily a reaction of the acidic flux residues with the basic material of the neutralizer. In addition to neutralization, chelating agents are added to prevent deposition of metallic salts on the assembly (Ellis 1981). The end products of the reactions are then rinsed off the assembly with clean water and the assembly is dried. The cleaning process is controlled by a monitoring of the pH of the neutralizer bath and replacing it when the supply of available base is reaching exhaustion.

Saponifiers.   Saponifiers derive their name from the word *sapo*, Latin for *soap*. A saponification reaction is the forming of a soap, i.e., organic salt, from the reaction of a base with an organic acid. Whereas most people are familiar with the soap derived from the hydrolysis reaction of animal or vegetable triglycerides with sodium or potassium hydroxide, the soap in question here is usually the result of the reaction between abietic acid (or one of its isomers) in rosin with an organic or inorganic base.

The organic based saponifiers are the most common. Typically these are an amine such as mono-, di-, or triethanolamine. The basic components of such a material are listed in Table 4.2.

Saponifiers are normally used at temperatures above room temperature to speed the reaction of the amines with the saponifiable portion of the rosin and reduce foaming.[5] As noted in Table 4.2 surfactants and organic solvents are sometimes

---

[5]Foaming is not just a function of temperature (although elevated temperatures reduce solution viscosity and allow foams to break faster). It is also a function of loading: the amount of flux residue in the wash tank. For rosin fluxes the foaming agent is the rosin soap formed in the cleaning reaction. If a water soluble flux is being used (with or without a saponifier being present in the wash section) surfactants from the flux, if allowed to concentrate in the wash section, may also cause foaming.

used to aid wetting and removal of nonsaponifiables, respectively. The latter is often necessary because all material in the rosin matrix is not saponifiable. (For a general description of the detergency process as related to surfactants and additives see Rosen (1978).

The reactive part of the cleaning process is followed by a rinsing process to remove the saponified rosin (the soap) and all ionic materials. Here the type and amount of surfactants in the saponifier mixture come into play. These surfactants aid in the cleaning of the assembly by lowering the surface tension of the water. As will be discussed later in this chapter, the use of these materials is two edged: they both aid in the cleaning of the flux residue and they may become a residue themselves on the final assembly.

Saponifiers are also manufactured with the reactive base being inorganic. Although some may balk at the use of an ionic material in the cleaning of an electronic assembly, it has been my experience that these materials are very effective and generally require no more rinsing than do their organic counterparts discussed above.

It should be noted that both these types of saponifiers are reactive not only with rosin flux residues but also with some metals, plastic parts, and people. Accordingly, care should be taken in their use. Manufacturers should be consulted as to the proper concentration needed and on the control of this concentration for proper cleaning without damage to sensitive parts of the assembly.[6] Before committing full scale production to the mercy of any cleaning process it is wise to test the effects on the assembly. (This point is discussed in a later section.) And for safety operators should be trained in the use of these materials and how to deal with spills and accidental sprays.

While saponifiers have been discussed here in reference to their use with rosin based fluxes they also enhance the cleaning of water soluble flux residues. This may be due to a "replacement" mechanism in which any adsorbed hydroscopic surfactant residues are displaced from the surface with the saponifier base (Zado 1983), or it may be due simply to the improved "wettability" of the cleaning solution with the added surfactants. Saponifiers also aid in the removal of processing oils, fingerprints, and other common soils. Their utility is not necessarily confined to use with rosin fluxes.

Semi-Aqueous Processing.   One of the best examples of a study showing the advantages of using both a solvent and water in the cleaning of high reliability

---

[6]Most manufacturers suggest concentrations in the 5–10% range. Concentration may be controlled by monitoring the pH of the wash solution. It can also be controlled through simple titration methods (Chung, Corsaro, and Jancuk 1983). My personal experience has been that once process flow is defined, replacement of the wash solution on regular intervals is the easiest method of maintaining concentration control. pH monitoring and the other chemical methods are then used as process checks.

assemblies is that done by Sanger and Johnson (1983). Of the variety of cleaning processes studied, the best method combined nonionic cleaning using solvents with ionic cleaning using water. While solvents are under something of a cloud today, the basic principles are still valid. Indeed, saponifiers really use a theoretically similar process that first attacks the water insoluble rosin and then removes the resulting soap and other ionic and water soluble materials.

Today, however, semi-aqueous processing generally refers to the use of a solvent (usually a mixture) that first removes nonionic materials and is then itself removed by a water rinse. The details of such processing are covered in Chapter 5. A well known example of such a process is terpene emulsion cleaning (Munie and Wenger 1988)

4.3.2. Equipment

No matter what cleaning medium is chosen for an aqueous process, there must be some method of getting it on and off the assembly. There are now numerous manufacturers of aqueous cleaning equipment for an assembler to choose from, so the reader will excuse the author if, to begin, I say I am not an expert on them all. Indeed, the number of features available on any given system is probably limited only by budgetary concerns. So what follows will be based on my experience in the aqueous cleaning field and should be treated as one person's view rather than a comprehensive treatise. From that perspective the basic methods, I believe, can be divided into two types: batch and inline.

Batch Cleaning Systems.   Batch cleaning systems can be roughly divided into those that use multiple stations for cleaning and those that perform cleaning in a single station.

Multistation cleaners are not a common off the shelf item for most manufacturers. Modified sinks are more common. Within AT&T, several custom made systems have been developed for small scale batch aqueous cleaning. The systems have used a series of separate tanks to perform the cleaning, rinsing, and drying sections of the process. Liquid agitation is provided by a spray system in the wash section. Rinsing is provided by a deionized water spray. Drying is an oven bake.

An example of a system in which all processing steps are done in a single enclosure is shown in Figure 4.1. Here the number of steps and duration of each step is at the discretion of the processing engineer. Thus a unique program can be set up for each product type. Agitation for the wash and rinse cycles is provided by a pumping system that irrigates the assemblies in a manner similar to a kitchen dishwasher. Drying is done by heated air flow through the enclosure at the end of the rinsing cycle.

Inline Cleaning Systems.   Inline systems are usually of the spray or underbrush

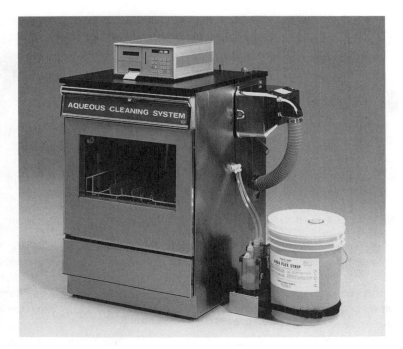

Fig. 4.1.   Batch aqueous washer. (Photo courtesy of ECD Inc.)

type. The application of ultrasonics with aqueous processing is a possible alternative that has not received as much acceptance as yet as the two aforementioned processes.

Spray Systems.   Inline spray cleaning systems are available from several manufacturers. They range from large units with multiple process stages to simpler units designed for space limited operations. An example of a large inline unit is shown in Figure 4.2.

As was described above the stages of these pass through operations are wash/rinse/dry or perhaps some multiple thereof. Figure 4.3 shows a simplified schematic. Vendors are often quite flexible in the design of such systems. Modular machines are available whose stages are customized as per request. All these allow a assembler to add as much cleaning or drying capacity in each stage of the process as he or she feels is necessary.

The use of modular design also allows an assembler to run a variety of process types on one machine. As has been described above aqueous cleaning can be done with water or water and some combination of chemicals matched to the cleaning needs of the flux/soldering process. (See Chapter 2 of this book for

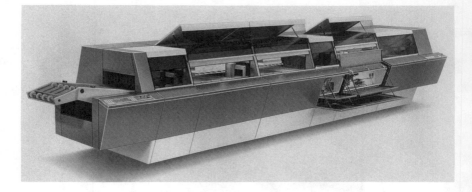

Fig. 4.2. Inline aqueous spray cleaner. (Photo courtesy of Electrovert Corporation.)

details on these considerations.) Thus, for one product, if a rosin flux is used a saponifier may be added to the first stage of the machine and water only used in subsequent sections (except the drying section!). If a water soluble flux should be used on another product the same machine may be adapted to a flow-forward cleaning process with water only.

As a note on the idea of using water only cleaning for water soluble flux I must add that from personal experience, the flow-forward system that many vendors advocate is extremely effective as a cleaning process. Pure water is added at the final rinse stage and cascades forward until the overflow from the first tank is discharged from the machine. An incoming assembly will then see a gross (and dirtiest) rinse first. Each succeeding cleaning section exposes the assembly to a rinse/wash process that is cleaner than the preceding section. The final rinse before dry is the last and cleanest part of the cleaning cycle.

Underbrush Systems.   Underbrush cleaning has been in use within the Bell System and postdivestiture AT&T for many years. (Habib 1983, Chung, Corsaro

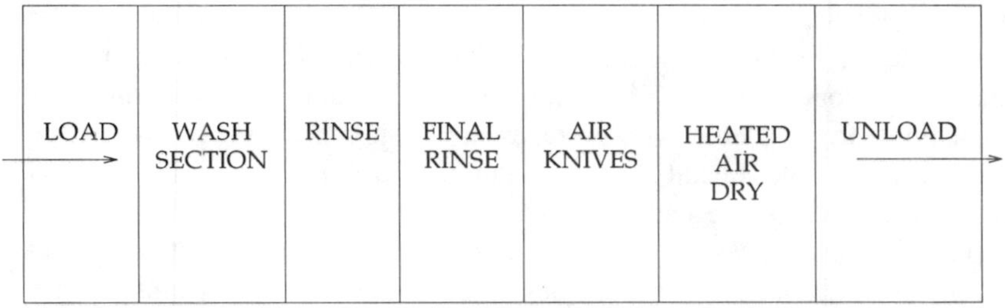

LOAD	WASH SECTION	RINSE	FINAL RINSE	AIR KNIVES	HEATED AIR DRY	UNLOAD

Fig. 4.3.   Schematic of an inline spray cleaner.

and Jancuk 1983, Sheldahl and Horton 1987). Washing and rinsing of the bottom side of the assembly are done by means of rotating brushes. The rotation of these brushes may be in a plane perpendicular (Chung, Corsaro, and Jancuk 1983) or parallel (Sheldahl and Horton 1987) to the bottom of the printed wiring assembly.

The underbrush process is applicable for through hole and top side reflow surface mount technology in which:

1. Testing is:
   A. From the bottom side of the assembly (bed-of-nails).
   B. Functionally through a connector.
2. The flux residue left on the topside is of a type that can remain uncleaned without affecting the reliability of the assembly.[7]

Although underbrush cleaning was originally done using chlorinated solvents, environmental concerns have caused a migration to aqueous saponifying detergents and terpenes (Chung, Corsaro, and Jancuk 1983, Munie and Wenger 1988, Sheldahl and Horton 1987). The advantage for an assembler who has designs compatible with this type of cleaning is that the components never directly contact the cleaning medium. It must be noted that materials qualification is still recommend to insure compatibility of the board with the flux/cleanin process. Suggestions for such qualification procedures will be discussed in subsequent sections.

Ultrasonic Cleaning. A promising technology that has not received general (or perhaps a better word is official) (Keeler 1988) acceptance in the cleaning of electronic assemblies is ultrasonic cleaning. Concerns as to the effects of ultrasonic agitation on the internal bonds of packaged ICs has caused most manufacturers to steer clear of this technology (Harmon 1974). However, this process has shown great promise in tests run at the Navy's EMPF (Paul 1988). It is possible that with proper design of assemblies and control of the process these problems will be overcome. Further studies are now in progress at EMPF to retest the reliability of this process. If concerns about the effect of ultrasonics on the internals of packaged ICs are answered this cleaning process may prove to be a boon to those desiring a rigorous cleaning of densely packed, low standoff, surface mount assemblies. Some European manufacturers are using ultrasonics in inline processes and apparently having good results (Hoffman 1989, Svenson

---

[7]Telecommuncation fluxes for the Bell System have typically been of this type. Specifications for such flux types are supplied to potential Bell operating company suppliers by Bell Communications Research (Bellcore).

1989). The use of ultrasonic cleaning is covered in detail in Chapter 5 of this book.

4.4   PROCESS QUALIFICATION AND CONTROL

By the term *process qualification* I mean the sequence of tests, experiments, judgments, and/or guesses by which one decides that a process will yield the intended end result. Process control is the feedback loop by which one monitors the process as it makes actual product and uses the results of that monitoring to adjust the process operation to insure that the process continues to deliver the intended results.

In the following sections the reader will learn nothing about the proper statistical sampling methods for controlling processes. Many of the specifics as to techniques for qualification will also be found wanting. What will be presented are strategies in accordance with prevailing wisdom on how process change and operation can be brought about and monitored, respectively. It is hoped that this information will be of use to an assembler considering aqueous technology for the cleaning of printed wiring assemblies.

4.4.1.   Qualification Strategies

There are two approaches that the author has seen applied in the qualification of process: materials qualification and assembly qualification. They are not mutually exclusive. Best results are obtained when the schemes are used complementarily.

Materials Qualification.   Materials qualification is dependent on some set of standards or limits that an assembler imposes on incoming materials in the assembly, flux, and other materials. The intent of these limitations is to insure that, upon combination, all of these materials have such a benign nature and interaction that the reliability of the product is insured without further testing.

An example of materials qualification can be found in ANSI/IPC-SF-818, "General Requirements for Electronic Soldering Fluxes." Another example in this area is the commonly used flux specification, MIL-P-14256 D. The intent of such a specification is to define a material's performance in a process by specific tests. These tests are assumed to adequately cover the types of environments that real product will see in operation. Thus the behavior of the various materials used during the assembly process is assessed beforehand. And the reliability of the assembly is supposedly insured by conformance to the tests.

Two problems with such procedures are that the tests do not accurately mirror either the assembly configuration or the field environment. In general I believe the errors have tended to be on the conservative side. Thus it is not so much that gross liabilities have been incurred by the use of such procedures as that a rigid methodology has been imposed on technology with little understanding as to how laboratory experience translates into manufacturing and field performance. Once such a test has been used on a particular process and results have been good, there is a tendency to assume that the test method is completely portable, i.e., applicable to all new technologies.

Assembly Qualification.   If an assembler has a good idea as to the type of environment the equipment will see, a strategy of assembly qualification can be implemented. As with the materials qualification described above, this would consist of testing assemblies manufactured in different process conditions, under some type of environmental stress to assess the effect of process on performance.

A very serious problem that I have seen in such techniques is the lack of concrete proof that a particular type of environmental stress translates into performance of a device under field conditions for any particular length of time. These problems can, I believe, be broken down into:

1. Lack of constant field conditions. Use, location, maintenance, and so forth are for the most part out of the hands of the assembler.
2. Insufficient knowledge as to how devices fail in the field. There are many more ways for assemblies to fail than there are parts in the assembly. Which one is the greatest contributor? Unless a careful study of field behavior and of the breakdown mode of failed assemblies is collected, a manufacturer will have no idea if process, materials, use conditions, or incoming part quality (to name but a few reasons) is responsible for assembly failure.
3. Difficulty in reproducing failure modes in the laboratory. Often assembly qualification involves use of "acceleration factors." These illusory numbers are the means by which laboratory failure rates are transformed into predictions of field lifetime.
4. Transformation of available knowledge about failure modes into a regimen of testing that accelerates the rate at which the relevant failure processes occur such that intelligent predictions can be made about the effect of process conditions on field reliability.

Please, after reading the above do not despair. There are methods for accelerated testing of assemblies that do provide assurance as to field performance. Military specifications and some telecommunications standards are good sources of information on methods for stressing assemblies. The general methods I believe can be grouped as follows:

1. Static environment acceleration. The assembly is exposed to temperature,

humidity, and electrical conditions that are in excess of those that are expected in the field. This test is often referred to as the THB (temperature, humidity, bias) test. The choice of conditions may be set by a customer's requirements or may be chosen judiciously by the assembler. In the latter case it is wise to pick conditions which induce aging without inducing failure modes that are unlikely to be seen by a device in the field, e.g., conductive anodic filamentation (CAF) in printing wiring boards.

2. Cyclic temperature environments. The assembly is cycled through high and low temperature conditions to induce stress. This is a basis for many military tests.

3. Contamination of the assembly: the assembly is deliberately exposed to extremes of dirt such as ionics or oils, e.g. salt spray. The resistance of the assembly to such conditions is expected to reflect on the assembly process reliability, e.g. the resistance of a conformal coat as it relates to the cleanliness of the board before coating.

My experience has been primarily with the first item. This has been because of the environments in which telecommunciations equipment functions. These environments tend to be static in temperature and humidity over long periods of time with little intrusion (except in special circumstances) by gross contamination. Accordingly the test regimens have reflected that fact. Therefore I would make the point that it is best to match (when customer requirements are not already set) test with use.

Qualification Summary. As I mentioned at the beginning of this section the best policy, especially where reliability considerations are of the highest concern, is a combination of materials and assembly qualification. In employing this combined approach try to look not just for conformance with requirement but for response of the test to a well designed variety of process conditions. (This will be dependent on how much time and resources an assembler has at his or her disposal.) By employing testing for overall response an assessment as to what is and is not important can be made. I mention this because it is important not just to "pass" any given test regimen but to understand why a particular set of process conditions can yield a particular response.

4.4.2. Cleanliness Evaluation Techniques

Within this section test methods will be discussed that deal with the evaluation of cleanliness. As such there will be some overlap with the above section on qualification. Indeed tests such as Surface Insulation Resistance (SIR) can be used for both materials qualification and for evaluation of the effectiveness of the cleaning process. None of the tests described will give an instantaneous real time evaluation of assembly cleanliness. All are end of the line checks. Accordingly, I feel that their greatest value lies not in their use for inspecting in quality but for providing information as to what parameters, e.g., flux and process time, are important in producing a clean, reliable product. They should be used as tools for the acquisition of knowledge as to what actually governs assembly cleanliness. It is that usefulness that I hold far above any value these techniques may have as inspection methods. Knowledge gained from such testing can be used in making the process and material choices that are so important to successful design of the electronic manufacturing process.

Analytical Methods.   Cleanliness of assemblies has been assessed by amounts of the two main types of residues: ionic and nonionic.

Ionic Residues.   Ionic residues can come from a number of sources such as flux activators, plating salts from board manufacture, fingerprints, or the breakdown products of rosin or surfactants. Accordingly the effect of the material on assembly reliability is also variable. Some methods of assessing contaminant levels are discussed below.

A good, reliable method of detecting the overall level of ionic contamination can be accomplished by the effect of these materials on the conductivity of extracting solutions (Brous 1979). This straightforward measurement technique has made evaluation of ionic cleanliness levels fairly easy.

An example of a commercially available instrument designed for quick assessment of ionic contamination is shown in Figure 4.4. Briefly the assembly or device to be tested is immersed in a circulating bath of deionized water and isopropyl alcohol. (The relative proportions of these materials can vary to some extent.) For the instrument shown, the solution is heated to achieve better extraction and more accurate assessment of total ionic contamination. The solution is then pumped through the cell containing the assembly, a conductivity meter that measure the amount of extracted ionic material, and (finally) deionizing columns that remove the ionic contaminant and prepare the solution for return to the extraction chamber. The test can be run either in the mode of total ionics extracted, (i.e., the process continues until no increase in conductivity of the deionized extractate occurs when it passes over the assembly) or for a set time with some limit to total ionic material extracted in that period of time being allowed by end point requirements. The former method is of value in assessment of a new process while the latter has value as a process check.

A question arises as to what any of the numbers obtained mean. Indeed, between different instrument suppliers there are different levels of sensitivity associated with the extractive ability of the different machine configurations (Ellis 1986). With newer test equipment one may suddenly be lead to believe that a previously satisfactory process has now gone awry. This may be due to

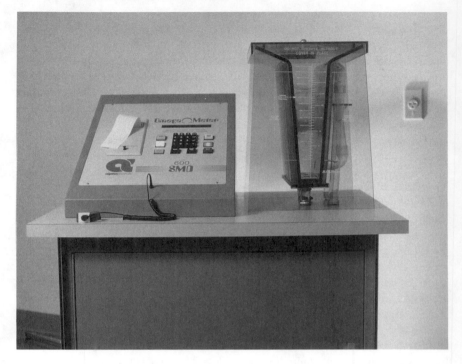

Fig. 4.4. Equipment for measuring ionic contamination of assemblies. (Photo courtesy of Alpha Metals Inc.)

nothing more than the increased sensitivity of the equipment. So although the contaminant level remains unchanged, the new equipment is simply able to measure more, raising fears that the assemblies have suddenly gotten dirtier (Carpenter 1989)! (The reader would do well to consider the effects of advances in analytical sensitivity in areas of environmental concern. In my opinion, our ability to measure lower and lower levels of 2,3,7,8-tetrachlorodibenzo-p-dioxin has done more to increase our fears over presence of this chemical than it has been able to increase our understanding of that chemical's effects.)

Exactly what is being measured cannot be directly determined by such equipment. Thus while machines are calibrated in terms of amount of sodium chloride equivalent that is present per unit area, they cannot assess what form the ionics are actually in e.g., free halide ions or rosin acids. In the former case there may be an immediate danger to product reliability. In the latter, concern may be extraneous to assembly reliability.

A method that is of use in quantifying the specific types of contaminants and thus providing information as to actual effect of the ionics present is ion chromatography (Wargotz 1979). While this method is more difficult than the general extraction techniques, it allows for a deeper understanding of the nature of the contaminants and the effect of process changes. The method, however, suffers from the disadvantage that it requires assessment beforehand as to the nature of the material to be analyzed for. In addition it cannot match the speed and convenience of the general extractive techniques.

Another example of ionic analysis which employed both extractive and spectroscopic methods for assessment of cleanliness in surface mount assembly is given in the reference by Wargotz, Guth, and Stroud (1987). I point this out both because of personal knowledge of the study and because the combination of techniques was useful in qualifying a previously untested process, i.e., wave soldering of surface mount devices with water soluble flux followed by water only cleaning.

As a note on the difficulty of correlating extractive techniques with actual species present, the tendency of some substances to remain immune to extractive analysis, and the potential for using spectroscopic methods the reader is referred to the work by Kenyon (1979b). This particular study, along with the work of Wargotz, Guth, and Stroud, (1987) shows the value of a combination of methods in dealing with cleaning problems related to ionic materials.[8] From personal experience I would advise that such extensive studies employing these more complex methods be reserved for problems of the specific type treated in the references cited: unusual cleaning difficulties and process qualification.

In summary, choosing between the above techniques for process monitoring and/or qualification must then be left to the readers discretion. For process monitoring the most viable choice will be the general extractive method. Process qualification on the other hand may call for more subtlety. All ions are not created equal in their effects on electronic assemblies. For a greater understanding of process effects more in-depth analytical techniques may be required.

Nonionic Residues.   Nonionic residues may be rosin or surfactants from fluxes or oils from soldering or handling. As such they too present a spectrum of possible effects on assembly performance.

Rosin residues have proved analyzable by extraction followed by spectrographic analysis (Archer, Cabelka, and Nalazak 1987, Kilma and Magid, 1987, Getty and Barrett 1989, Wittmer and Boomer 1989) These techniques have used both straight extraction and spectroscopic measurement, as well as extraction followed by chromatographic separation.

Another and more qualitative method for non-ionics is the extraction by ace-

---

[8]I say ionic materials because the residue examined by Kenyon was metal based. The reader should note that this "ionic" material was not ionizable by extractive techniques! As such it required the more sophisticated approach outlined in that paper.

tonitrile. This method is described in the IPC test methods (IPC 1984). The assembly is washed with acetonitrile and the resulting solution allowed to evaporate on a glass slide. The slide is then visually or spectroscopically examined for the presence of a residue. The disadvantage of such a technique is its lack of quantitativeness. The advantage of the technique is its simplicity and its ability to identify different types of residues through the Infrared analysis of the slide residues. As such it can be used as a screening technique that is employed before more specific analysis for rosin or surfactants is tried.

Surfactants used in water soluble fluxes and saponifiers have tended to be of the nonionic type. This limits their detectability by the extractive techniques described above for ionics. High pressure liquid chromatography (HPLC) has been successfully used to detect these nonionic residues (Chung and Eagle 1981).

Other methods have been used to investigate nonionic residues in assemblies. Some of this work has been done in connection with the testing of nonchloro-fluorocarbon (CFC) cleaning processes (Getty and Barrett 1989). In this instance mass spectroscopy was used to investigate postcleaning residues.

Limitations on the Methods. In summary I would like to comment that knowledge of a substance's presence on an assembly is a two edged sword. It is nice to know what is on the board, but a knowledge of how that material (and at what levels!) affects the assemblies' performance and reliability is also needed. And this in turn requires some understanding of both the limits of the analytical technique and the behavior of the assembly and contaminants under the conditions of actual field operation. In closing I cite some examples for the readers consideration:

1. Rosin residues will be found at some level on every board soldered with rosin flux. What is the effect of that residue? Within the telecommunications industry safety of topside residues of flux meeting the telecommunications flux standards has been (I believe) adequately demonstrated. However, for military equipment undergoing conformal coating such residues can be disaster (Tautscher 1976).
2. White residues on boards have been known for some time (EMPF 1988). In most cases the residues have had only a cosmetic effect. In some instances the presence of such material has affected assembly performance. (Kenyon 1979b).
3. Surfactant residues have been found on assemblies processed with water soluble fluxes and cleaned with aqueous detergents. What is the effect of these residues? While tests have shown that such materials affect the surface insulation resistance of boards (Brous 1981), the reduced resistance does not show the similar types of failures associated with ionic contamination (Zado 1983).

In each case something can be found on a board with one of the above mentioned analytical techniques. The effect of that "something" may vary considerably depending on what the board is and where it operates. Intelligent use of any technique demands that numbers produced by that technique be understood and not just compared to a specification value.

Electrical Testing.   The procedures described here are applicable to both the assessment of cleaning process efficiency and the qualification of that process. The two most common are the surface insulation resistance test and the electromigration test.

Surface Insulation Resistance (SIR).   Surface insulation resistance (SIR) is the measurement of the insulating properties of an assembly substrate. This may be done on unprocessed boards with or without solder mask to test the inherent properties of a substrate. Examples of such test procedures can be found in documents such as IPC-TM-650, "Test Methods Manual" or Bellcore's TR-TSY-000078, "Generic Physical Design Requirements For Telecommunications Products and Equipment." It may also be done to assess the properties of flux. Examples of this can be found in Chapter 2 of this text and in ANSI/IPC-SF-818, "General Requirements For Electronic Soldering Fluxes." When used as an assessment of cleanliness the method can detect both ionic (Chan and Shankoff 1988) and nonionic (Brous, 1981, Zado 1983) residues. An example of a commercially available measuring system is shown in Figure 4.5.

Associated with this test method is a variety of test patterns. Examples of such patterns are the AT&T broad line pattern, the Bellcore stripped pattern, the Military Y, and the IPC B-25 pattern. Each of these test vehicles represents a different spacing between conductors and as such will show different sensitivity and repeatability in testing. (Gorondy 1988).

It is not the intent of this text to discuss all the considerations involved in the measurement of surface insulation resistance. Details associated with the measurements are covered in Ellis (1986) and Gorondy (1984, 1985, 1988), (Wiser heads than mine have mulled over the issues associated with SIR measurements!) Still, based on personal experience, there are some points that I feel are particularly relevant where aqueous processing is concerned.

1. Although failure modes have been postulated and charge to failure values determined (Boddy et al. 1976, Delaney and Lahti, 1976), much of this information may not be directly applicable to the presence of surfactants that results from an aqueous process (Zado, 1983). This in no way implies that SIR should be considered valueless as applied to this new technology. But it must be understood that a simple readout of a value (Zado, 1983) may not accurately predict assembly performance.

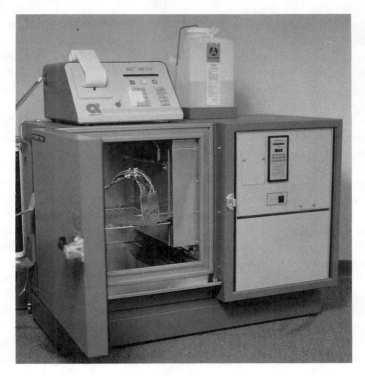

Fig. 4.5. Equipment for the measurement of surface insulation resistance. (Photo courtesy of Alpha Metals Inc.)

2. Just as surfactant residues produce a lowering of SIR, rosin residues, even in a highly activated system, can produce very high SIR depending on the amount of rosin adsorbed on the surface of the assembly (EMPF 1989).
3. As a comparative method that is coupled with other analytical techniques and process qualification methods, SIR is extremely useful. In the sense that it provides an exact yes/no answer (as is sometimes expected of other techniques) it is not infallible.[9]
4. The techniques for SIR measurement require great care. Slight changes in process conditions, humidity variations within the test chamber, fluctuations of the measuring equipment, or any of a number of factors can influence the repeatability and accuracy of SIR testing. (The references by Gorondy (1984, 1985, 1988) are very illuminating at this point. The report by Maguire (1989) is also very useful.)

---

[9]Indeed, in the sense that it does not always give difinitive answers it is more like the riddle spouting oracle at Delphi.

5. The relationship of SIR to reliability may not be exact. Work on this topic (Chan and Shankoff 1988) has served to clarify some issues. I would, however, venture the opinion that more needs to be done in this area.

The reader should not conclude that I have no use for SIR techniques. On the contrary, they have been and continue to be useful tools in my work on flux and cleaning issues. But, having been burned myself, I would be lax if I did not warn others that such measurements are but part of the total set of tools needed to evaluate assembly processes.

Electromigration Test.   A second test that requires a less elaborate set up than the SIR test is the electromigration test. The details of this test for assessing board cleanliness are outlined in IPC-TM-650.

The test involves exposing a standard conductor pattern to a given process and testing for the flow of current in the presence of high voltage and water. Some ionic materials react electrochemically. This results in the migration of metal ions from the conductor with accompanying dendritic growth. The time for such growth is monitored by changes in current flow between the conductors. Rapid growth of the dendrites or high current flows is considered indicative of a poor cleaning process for the flux being used.

As was mentioned above, the test requires less equipment than the SIR test. However, because of a lack of concensus with respect to interpretation of test results it has not been as popular in its application to cleaning problems as has SIR.

Summary.   All of the tests mentioned above allow the assembler to assess the effectiveness of a process in producing clean, reliable electronic assemblies. All of these tests allow checks for cleanliness at the end of the process. Some of the checks such as the Acetonitrile Extraction test are relatively straight forward. Others such as the SIR involve more elaborate test methods. My opinion is that a combination of the above described tests is best for process qualification. Such a combination gives the assembler the data needed to make intelligent choices of process conditions and materials. After process qualification and prove-in, and as confidence on assembly performance is gained, these tests can take the role of process checks. Some may then be eliminated, with others applied only on sampling basis.

4.5.   ENVIRONMENTAL CONCERNS

On the scale that modern society lives there is probably no completely neutral process from an environmental standpoint. Perhaps this is due to the fact that we are part of the environment and our presence will always, in some way, affect other parts of the system. But just as we try to control undesirable social behavior, for both moral and pragmatic reasons we should try to control our environmental behavior. If, without eliminating society, we cannot eliminate all our effects, we should at least reduce as much as possible the undesirable influences we do exert on the environment.

I claim no special knowledge of environmental concerns either regulatory or technical. From a practical standpoint the problems I have run into related to aqueous cleaning have been those of heavy metal contamination, pH, and biochemical oxygen demand (BOD) of the effluent. Accordingly I will mention these briefly and close with some comments on how such issues fit in with the other topics discussed above.

4.5.1. HEAVY METALS

Dissolved metals in the effluent from an aqueous cleaner are a serious concern. These metals are mainly lead, copper, and tin. To a great extent their concentration is determined by process conditions (Woodgate 1983, UNEP 1989). Treatment of process before discharge may be necessary for their removal (Bernstein 1981).

However, a first step to take at the machine discharge (before any treatment) is filtration. Solder particulates that otherwise will end up in treatment can easily be removed with filtration. (Note: The filtered material and the filter itself need be disposed of in a safe and approved manner.)

A note of caution: It has been my experience that any use of cleaning materials that can react with the either the dissolved metals, i.e., soluble ions, or with the solid metals on the assembly itself can complicate the treatment process. Examples of such complicating materials are amine saponifiers that can complex with the metal ions and chelating agents that may be added to neutralizers. Accordingly, when addressing the problem of metals in effluents make sure that all chemical players in the process are taken into consideration before settling on a treatment scheme.

4.5.2. pH

The pH of the discharge can also be an environmental problem. This is most common when saponifiers are used. But monitoring of this parameter should be done in all cases. The most common method of treatment is neutralization. An automated system for neutralization is described in the paper by Sheldahl and Horton (1987).

4.5.3. BOD

Biochemical Oxygen Demand or BOD is the amount of oxygen required by aerobic organisms to metabolize the organic materials in a wastewater solution.

A high BOD Is generally reflective of a large concentration of organic material. (The correlation is not direct because of the different metabolic pathways involved in the breakdown of different compounds.) Discharge of high BOD effluents into natural waterways will deplete the available oxygen in the water as the breakdown of the material proceeds. Limits are therefore set on the BOD (in units of milligrams of oxygen per liter) of the effluent discharged into any waterway. Treatment required to meet these limits may be necessary depending on the nature of the process.

4.5.4. Summary

The above is only a brief description of three of the environmental problems associated with aqueous processing. No exact formula has been presented by which the reader can jump to complete environmental compliance for an aqueous process. Such a formula does not exist (at least universally) because of the differences in federal, state, and local laws on water quality and treatment. And this point brings us back to the issue raised previously on design: up-front consideration of these concerns, along with things such as flux and machine choice, will make the difference between success and failure. At the beginning of the design process environmental authorities, both internal and external to the operation, should be consulted. The local regulatory agencies can supply details as to what discharge quality is required. The process engineer who works with his organization's environmental group to meet these requirements will have fewer problems in the start up and running of the process. The engineer who ignores this may find that he or she is involved in an endless game of regulatory catchup.

REFERENCES

ANSI/IPC-SF-818, "General Requirements For Electronic Soldering Fluxes." 1988.

Archer, W.L., Cabalka, T.D., and Nalazak, J.J. 1987. "Quantitative Determination Of Rosin Residues On Cleaned Electronic Assemblies." 11th Annual Electronics Manufacturing Seminar, China Lake, California, February 18–20.

Askengren, L. 1989. "Solvent Cleaning or Aqueous—Energy Consumption and Cleaning Costs." Technical Report From Surface Engineering AB, Lerum, Sweden, May 16.

Aspandiar, R., Piyarali, A., and Prasad, R. "Is OA OK?" *Circuits Manufacturing,* **26** (4), pp. 29–36 (1986).

Bahr, K.E., and Andrus, J.J. 1989. "A Kinetic Approach to SMD Aqueous Cleaning." *Circuits Manufacturing,* **29** (11), 52–56.

Bernstein, S. 1981. "Mass Soldering and Aqueous Cleaning Process Operated In Compliance With Water Pollution Regulations." *NEPCON Proceedings.*

Boddy, P.J., Delaney, R.H., Lahti, J.N., Landry, E.F., and Restrick, R.C. 1976. "Accelerated Life Testing Of Flexible Printed Circuits, Part I—Test Program and Typical Results." *IEEE Reliability Physics Symposium Proceedings,* **14**, 108.

Brous, J. 1979. "Extraction Methods For Measurements Of Ionic Surface Contamination." *Surface Contamination: Genesis, Detection, Control,* Vol. 2, K.L. Mittal, Ed. New York: Plenum Press.

Brous, J. 1981. "Water Soluble Flux and Its Effect on P.C. Board Surface Insulation Resistance." *Electronic Packaging and Production,* **21** (7), 80.

Brous, J., and Schneider, A. 1984. "Cleaning Surface Mount Assemblies With Azeotropic Solvent Mixtures." *Electri-onics,* **30** (2).

Cabelka, T.D., Archer, W.L., and Trombka, J.A. 1986. "Cleaning SMCs With Chlorinated Solvents." *Proceedings of NEPCON West, Anaheim, California.*

Carpenter, R. 1989. Private communication, Alpha Metals.

Chan, A.S.L., and Shankoff, T.A. 1988. "A Correlation Between Surface Insulation Resistance and Solvent Extract Conductivity." *Circuit World,* **14** (4) 23–26.

Chung, B.C., Corsaro, V.A., and Jancuk, W.A. 1983. "Aqueous Detergent For Removing Rosin Fluxes." *The Western Electric Engineer,* **27** (1), 62.

Chung, B.C., and Eagle, J.G. 1981. *Proceedings of the IEPS Meeting, Cleveland, November.*

Comerford, M. 1984. "Cleaning PWAs: The Effect of Surface Mounted Components." *Electri-onics,* **30** (11).

Danford, A.L., and Gallagher, P.A. 1987. "SMD Cleanliness in an Aqueous Process." *Proceedings of NEPCON West, Anaheim, California.*

Delaney, R.H., and Lahti, J.N. 1976. "Accelerated Life Testing of Flexible Printed Circuits, Part II—Failure Modes in Flexible Printed Circuits Coated With U.V. Cured Rosin." *IEEE Reliability Physics Symposium Proceedings,* **14** 114.

Elliot, D.A. 1990. "Is Aqueous the Answer?" *Proceedings of NEPCON West, Anaheim, California.*

Ellis, B.N. 1981. "The Neutralization of Acid Fluxes for Soft Soldering." *Brazing and Soldering,* (1) 21.

Ellis, B.N. 1986. "Cleaning and Contamination of Electronic Assemblies."

Ayr, Scotland: Electrochemical Publications Ltd.,

EMPF. 1988. "An Organized Approach to Solving White Residue Problems," workshop sponsored by the Electronic Manufacturing Productivity Facility (EMPF) and presented at the IPC 31st Annual Meeting, Hollywood, Florida, April 18–22.

EMPF. 1989. "Ad Hoc Solvents Working Group Cleaning and Cleanliness Test Program: A Joint EPA/DOD/IPC Effort. Benchmark Test Results." Presented at the IPC 32nd Annual Meeting, Lake Buena Vista, Florida, April 22–28.

Getty, H., and Barrett, T. 1989. "Quantitative Non-Ionic Cleanliness Measurement by HPLC." IPC Technical Report TP-796.

Goldstein, I. 1980. "Aqueous Cleaning: A Systems Approach." *Insulation/Circuits,* May.

Gorondy, E.J. "Surface Insulation Resistance (SIR)—Part I: Development of an Automatic SIR Measurement Technique." IPC Technical Report TP-518. 1984

Gorondy, E.J. "Surface/Moisture Insulation Resistance (SIR/MIR)—Part II: Exploring the Correlation Between the IPC, Standard Industry, and Military Test Patterns—A Status Report." IPC Technical Report TP-543. 1985

Gorondy, E.J. 1988. "Surface/Moisture Insulation Resistance (SIR/MIR)—Part III: Analysis of the Effects of Test Parameters and Environmental Conditions on Test Results." Presented at the 33rd IPC Annual Meeting, Anaheim, California, October 24–28.

Habib, J.M. 1983. "Aqueous Underbrush Cleaning." *Proceedings of NEPCON East.*

Harmon, G.G. 1974. "Metallurgical Failure Modes of Wire Bonds." *IEEE Reliability Physics Symposium Proceedings,* **14** 131.

Hoffman, M. 1989. "Cleaning of Printed Circuit Boards with Isopropyl Alcohol." *Proceedings of the Conference on Countering of CFC Use Within the Electronics Industries, June 6–8, Stockholm Sweden.*

ICF Inc. 1987. "Use and Substitutes Analysis of Chloroflurocarbons in the Electronics Industry." Report to the U.S. EPA.

IPC-AC-62. 1984. "Post Solder Aqueous Cleaning Handbook." IPC, 7380 N. Lincoln Ave, Lincolnwood, Illinois 60646.

Keeler, R. 1988. "Ultrasonic Defluxing: Waves of the Future?" *Electronic Packaging and Production,* **28** (3), 52–54.

Kenyon, W.G. 1978. "Part 2—Vapor Defluxing Systems that Meet Today's PCB Cleaning Needs." *Insulation/Circuits,* **24** (3) 35–40.

Kenyon, W.G. 1979a. "New Ways to Select and Use Defluxing Solvents." *Proceedings of NEPCON West, Anaheim California.*

Kenyon, W.G. 1979b. In *Surface Contamination: Genesis Detection, and Control,* Vol. 2, K.L. Mittal, Ed. New York: Plenum Press.

Kilma, R.F., and Magid, H. 1987. "Analysis of Fluxes and Extracts Of Printed Wiring by High Performance Liquid Chromatography." *Proceedings of NEPCON West, Anaheim, California.*

Maguire, J. 1989. "Analysis of the Effects of Cycling versus Standing Humidity Environments and Electrical Test Parameters on Surface Insulation Resistance." IPC Technical Report TP-801.

Manko, H.H. 1986. "Soldering Handbook for Printed Circuits and Surface Mounting. New York: Van Nostrand Reinhold Inc. 1986.

Manko, H.H. and Ayers, L. 1989. "Panel Discussion: Aqueous Defluxing For Military Electronics." *Proceedings of NEPCON East, Boston, Massachusetts.*

Menon, V.B., Donovan, R.P., Michaels, L.D., and Ensor, D.S., 1988. "Evaluation of Various Solvents for Removal of Particulate Contaminants from Semiconductor Surfaces." *Proceedings of the CFC-113 Replacement Workshop, UCLA Engineering Research Center, Los Angeles, California, November 10.*

Munie, G.C., and Wenger, G.M. 1988. "Defluxing Using Terpene Hydrocarbon Solvents." Presented at the 31st Annual IPC Meeting, Hollywood, Florida, April 18–22..

Munie, G.C., Owens, S.L., Shaap, B.H., and Wenger, G.M. "Comparison of Cost for Aqueous Processes" Proceedings of CFC Alternatives Conference San Francisco CA 6/26/90.

Osterman, H.F. 1986. "How to Compare PC Cleaning Costs." Allied Chemical Information Bulletin, Reprinted from *Circuits Manufacturing.*

Paul, J.T. 1988. "Ultrasonic Cleaning—Can Current Technology Meet the SMT Challenge?" IPC Technical Report TP-798.

Rosen, M.J. 1978. *"Surfactants and Interfacial Phenomena."* New York: John Wiley and Sons.

Sanger, D., and Johnson, K. 1983. "A Study of Solvent and Aqueous Cleaning of Fluxes." Naval Weapons Center, China Lake, California, NWC TP 6427, February.

Sheldahl, S.A., and Horton, J.S. 1987. "An Environmentally Preferred Alternative to the Use of Chlorinated Solvents for Circuit Pack Underbrush Cleaning." IPC Technical Paper TP-625. Presented at the IPC 30th Annual Meeting, March 29–April 3, Atlanta, Georgia.

Svenson, J. 1989. "A Strategy to Eliminate the Use of CFCs as Cleaning Agents for PBAs." *Proceedings of the Conference on Countering of CFC Use Within the Electronics Industries, June 6–8, Stockholm, Sweden.*

Tautscher, C.J. 1976. "The Contamination of Printed Wiring Boards and Assemblies." Omega Scientific Services.

Trieber, J. 1989. "The Theoretical Aspects of Water vs Solvent Cleaning: Surface Tension, Wetting Capillarity, SMD Applications." *SMT/TAB News,* 3 (4).

UNEP Solvents Technical Options Committee. 1989. "Electronics Cleaning, Degreasing, and Dry Cleaning Solvents Technical Options Report."

Wang, A. 1987. "Effects of Wetting and Capillary Action on the Cleaning of Surface Mount Assemblies." IPC Technical Report TP-656.

Wargotz, W.B. 1979. "Ion-Chromatography—Quantification of Contaminant Ions in Water Extracts of Printed Wiring." *Surface Contamination: Genesis Detection, and Control,* Vol. 2, K.L. Mittal, Ed. New York: Plenum Press.

Wargotz, W.B., Guth, L.A., and Stroud, C.L. 1987. "Quantification of Cleanliness Beneath Surface Mounted Discretes Assembled by Wave Soldering of Printed Wiring." *Proceedings of Printed Circuit World IV, Tokyo, Japan, June 2–5.*

Wittmer, P., and Boomer, B. 1989. "Rationale and Methodology for a Standard Residual Rosin Test by UV-Visible Spectrophotometric Methods." IPC Technical Report TP-791.

Woodgate, R., Rahn, A., and Down, W.H. 1980. "Drying PWBs After Cleaning." IPC Technical Report IPC-TR-343.

Woodgate, R. 1983. *Proceedings of NEPCON East, Boston, Massachusetts.*

Zado, F.M. 1983. "Effects of Non-Ionic Water Soluble Flux Residues." *The Western Electric Engineer,* **27** (1), 40.

5
Alternate Defluxing:
Materials, Processes, and Equipment

Michael E. Hayes
Petroferm, Inc.

Kathi Johnson
Hexacon Electric Co.

5.1. INTRODUCTION

This chapter concerns alternate defluxing materials, processes and equipment. A functional definition of alternate defluxing techniques would describe alternate defluxing as any method which does not use halogenated solvents. In this chapter, the discussion is limited to those chlorine-free materials which are applied in what has come to be known as a semi-aqueous cleaning process.

Since their first description in the literature in 1987, semi-aqueous processes have been increasingly embraced as the most viable alternative to chlorofluorocarbons (CFCs) for cleaning printed wiring assemblies (PWAs). Several semi-aqueous materials have now appeared, and there is every reason to expect that additional candidates will continue to be identified. In this chapter, we will consider four semi-aqueous cleaners in some detail. It was our intention to provide information on each of the semi-aqueous products for which their manufacturers would reveal enough information to allow meaningful coverage. Adequate information proved hard to come by, so, unfortunately, details are lacking regarding the characteristics, use, and performance of some of these products. Nonetheless, it is our judgment that even this limited information will be a welcome addition to the knowledge base of most assembly personnel.

A semi-aqueous process can be defined as one in which cleaning is conducted in two steps. The first step is a solvent wash, and the second step is an aqueous rinse. The rinse step is necessary because, unlike CFC's or other halogenated solvents, the solvents commonly used in semi-aqueous processes are not volatile, so they will not evaporate. They must be removed by rinsing.

In years past the selection of a cleaning material was a fairly simple one. But in these times of environmental awareness we have come to realize that each person has a responsibility to fully understand the materials that they work with. This understanding is not limited to the specific process but more widely applies to the material used, its manufacturing process, toxicity, biodegradability, and impact on future generations.

In discussing the desirable attributes of a cleaning material it becomes important to describe some of the techniques for testing and classification of these materials. We have taken some time to describe these features in the environmental acceptability section.

5.2.   METHODS AND EQUIPMENT

In the sections that follow, we will attempt to provide an overview, along with a reasonable amount of detail where it seems warranted, on alternative defluxing materials, their properties and characteristics, and the processes and equipment in which they are used. Most of our attention will be focused on semi-aqueous cleaning, because that process is the recommended method for all five of the alternative defluxing materials about which enough information was available as of this writing to justify their inclusion in this chapter. First, though, we will review acoustic cleaning, commonly referred to as ultrasonic cleaning. Ultrasonics can, of course, be used with traditional defluxing materials. The process is included here because it is commercially important, and this chapter seemed the best place for its consideration.

5.2.1.   Acoustical (Ultrasonic) Cleaning

The use of ultrasonics in cleaning applications is widespread crossing into many different industries. Ultrasonic cleaning can be defined as the use of high frequency sound waves transferred to a liquid medium to produce cavitation. The ultrasound region is defined as the frequency range above the human hearing range (20 kHz). Acoustic cleaning encompasses both cavitational acoustics and noncavitational acoustics. The more traditional ultrasonic cleaning utilizes cavitational acoustic energy in the range of 20–99 kHz. Non-cavitational acoustic energy uses frequencies above 100 kHz and is also known as megasonics.

Cavitation is the phenomena that provides the mechanical action to remove soil from the workpiece being cleaned. Figure 5.1 shows an illustration of how ultrasonic cleaning works. Cavitation occurs as the sound waves act upon microscopic sized dissolved gas bubbles in the liquid medium. The ultrasonic energy causes the gas bubbles to expand and contract with alternating current. The

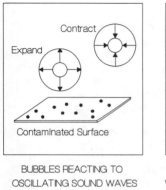

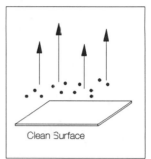

BUBBLES REACTING TO
OSCILLATING SOUND WAVES

BUBBLES IMPLODE; PRESSURE
WAVES REACH CONTAMINATION
ON SURFACE AND ATTACK
ADHESION OF PARTICLE
ON SURFACE

PARTICLES REMOVED FROM
SURFACE; SOLVENT HELPS TO
PREVENT REDEPOSITION

Fig. 5.1.   How ultrasonic cleaning works to clean a surface.

bubbles begin to resonate and at a critical point the bubbles will implode. As the bubbles implode tremendous forces are generated. This force is the mechanical action that is required to break the bonds between a particle of contamination and the workpiece being cleaned.

There are actually three different phases to the ultrasonic cleaning process. Each time an ultrasonic process is used the same three phases must occur for effective cleaning and therefor cavitation to occur. The first phase is the *degas phase*. This is where relatively large (above 60 microns)[1] dissolved gas bubbles are forced to the surface of the liquid medium. The degassing step is critical to the effective use of ultrasonic cleaning. If no degassing occurs the large bubble of dissolved gas would absorb the ultrasonic energy and impede cleaning.

The next phase is known as *gaseous cavitation*. This is where the smaller gas bubbles begin to expand and contract as exposed to the alternating sound waves. Some have thought that during this phase the liquid medium is effectively pumping onto the workpiece. The last phase of ultrasonics and the most critical is known as vaporous cavitation. This is where the expanding and contracting bubbles reach a resonant frequency and implode releasing pressure waves that have been recorded exceeding 1000 atmospheres of pressure.[2]

Transducers Types.   There are two different types of transducers which are used to produce ultrasonic frequencies: magnetostrictive and piezoelectric. Magnetostrictive transducers are based upon the property of a material to shrink and expand alternately when exposed to changes in electromagnetic fields. Magnetostrictive transducers operate typically in the 25 kHz range. Piezoelectric trans-

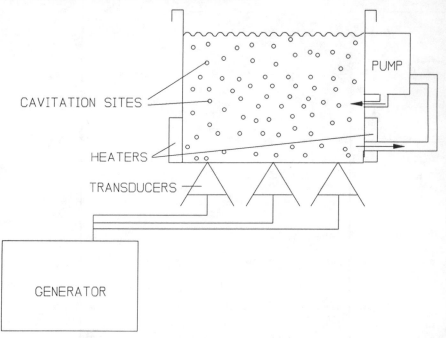

Fig. 5.2.   Schematic of a batch ultrasonic cleaning tank.

ducers are based upon the property of certain crystalline materials to shrink and expand when variable voltage is applied. Piezoelectric transducers used for ultrasonics cleaning operate typically in the 20–50 kHz range but are most commonly used at 40–50 kHz. In current ultrasonic cleaning equipment design piezoelectric transducers are more common. Figure 5.2 shows a schematic of a batch aqueous ultrasonic cleaning unit.

There are several factors that can have an impact on the success of an ultrasonic cleaning process; frequency, power level, solvent medium, and orientation of the workpiece. The selection of a proper frequency influences the cavitational intensity in the cleaning tank. The level of power selected will determine the distribution of cavitation in the tank. The selection of the optimum solvent medium can increase solvency for the contamination being removed. Orientation of the PWA in the cleaning tank can greatly affect the efficiency of the cleaning operation.

Frequency.   The pressure released when a resonating bubble implodes varies according to the frequency selected. This parameter is known as cavitational intensity. Lower frequencies (25 kHz) are said to generate higher cavitational intensities. The advantages of a higher cavitational intensity is offset by other negative parameters such as larger bubble size which limits penetration into small crevices and fewer cavitation sites per unit area. Higher frequencies (40–50 kHz) have lower cavitational intensities but have several other advantages. The distribution of cavitation sites is more uniform at this frequency range and the smaller bubble size allow for good crevice penetration.

Many people have tried to measure cavitation intensity in an ultrasonic cleaning tank with limited success. One of the easiest techniques involves the use of a sheet of aluminum foil immersed vertically into the cleaning medium for a specific amount of time. As cavitation occurs it actually erodes the thin aluminum foil. If the aluminum foil is held up to the light it is easy to observe the damage as seen by the holes in the aluminum foil. One of the problems with this technique is that it is difficult to maintain the foil in a single plane within the cleaning tank. If repeated foil test are performed it is very difficult to achieve any test consistency. Moderate success has been achieved when stiffening plates are mounted on the top and bottom two inches of the foil. These stiffening plates act to stabilize the position of the foil in the tank so that repeatable results can be obtained.

Other techniques for measuring cavitation are available. Cavitation probes have been available for many years. The key to successful use of these probes is to understand their limitations. One particular probe design, manufactured by NTK, incorporates a piezoelectric crystal to monitor voltage change which is then equated to frequency. The limits of this design are that it only measures peak voltage at one point in time, it does not discriminate between different frequencies and that the piezoelectric crystal are temperature sensitive. In their studies the Electronics Manufacturing Productivity Facility (EMPF) attempted to offset some of these limitations by linking a spectrum analyzer to the cavitation probe.

The spectrum analyzer output allowed for multiple sampling capabilities and was able to discriminate frequency and determine amplitude at each frequency. Amplitude was equated to power level. When this test was performed using a consistent plane within the tank it was possible to examine cavitational intensity and power in different areas of an ultrasonic tank. This technique was used to evaluate some of the commercially available ultrasonic cleaning equipment in EMPF's study on the use of ultrasonic cleaning for military electronics.[3] At the time of writing this project was still being completed.

Power.   There is no optimum power level that can be defined generally for cleaning because of the influences of equipment and transducer design. Bulat does however, specify that 1/3 watt per square centimeter is required to initiate cavitation.[4] It is important that the power is evenly distributed in the tank. If not, hot spots or dead spots can occur and effect cleaning performance. EMPF studies show that if the temperature in the ultrasonic tank is reduced than an increase in power is required to achieve the same level of cleanliness.[5]

Cavitation decreases as the boiling point of the solvent increases because dissolved gas bubbles are forced to the surface of the liquid and fewer sites for cavitation are available. Although solvency is increased at boiling temperatures it is unlikely that vaporous cavitation will occur. Many have claimed increased efficiency with the use of boiling ultrasonics but this work has not been well documented in the literature.

Selection of Cleaning Medium.   The selection of the proper cleaning medium is another important parameter to the success of ultrasonic cleaning. Obviously, the better a solvents capability to dissolve contamination the greater the cleaning performance. There are other features of a solvent as it relates to its use in ultrasonic cleaning that may not be very obvious. Bulat states that high vapor pressure liquids do not cavitate well. It is also known that high surface tension liquids release greater amounts of energy upon implosion.[6] This is why water is such an effective solvent for cavitation. It has both low vapor pressure and high surface tension. Surfactants may be added to the water if lower surface tension is desired but rinsing is critical if surfactants are used.

Orientation.   It is important that the workpiece being cleaned be perpendicular to the source of the frequency or the transducer. This orientation allows the maximum benefit from the sound waves. Some manufacturers locate their transducers on the side of the cleaning tanks but perhaps the best location is on the underside of the bottom of the cleaning tank (see Figure 5.2). As sound waves move through the liquid they bounce off the tank walls and the surface of the liquid. It is also important to hold the work piece in a rack or basket made of a material that will not absorb the ultrasonic frequency. Bulat shows that there is a 25–50% loss in cavitation when a polyethylene beaker is used instead of a glass beaker.[7] A good rule of thumb is to avoid any high density material such as a fine wire mesh basket because it will inhibit cavitation.

One of the advantages to using higher frequencies is the decreasing thicknesses of both viscous and thermal boundary layers. When these boundary layers exist, contamination is insulated from any acoustic energy. If the acoustic energy can't reach the surface because of these boundary layers then reduced cleaning efficiency occurs. At 30 kHz the viscous boundary layer is 4.4 microns and the thermal boundary layer is 1.1 microns. At 900 kHz the viscous boundary layer is reduced to 0.8 microns.[8] Figure 5.3 shows the relationship between frequency and boundary layer thickness. This property has been used in the design of a new high frequency cleaner from Branson Ultrasonic Corp. The system, known as MicroCoustic (TM), uses a 400 kHz acoustic array which continuously provides a narrowly focused field of compact wave fronts that effectively pump

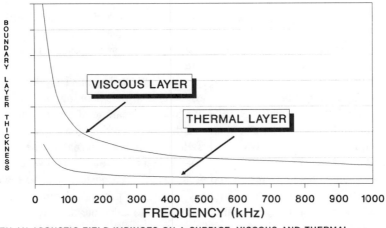

WHEN AN ACOUSTIC FIELD IMPINGES ON A SURFACE, VISCOUS AND THERMAL
BOUNDARY LAYERS EXIST. SOIL IN THESE LAYERS WILL NOT BE AFFECTED BY THE
ACOUSTIC ENERGY UNLESS THE FREQUENCY IS INCREASED.

Fig. 5.3. Relationship between frequency and boundary layer thickness.

Fig. 5.4. Photo of Branson MicroCoustic system. (Photo courtesy Branson Ultrasonics
Corp., Danbury, CT.)

solution at a high velocity beneath components. See Figure 5.4. Unlike most sprays which dissipate upon impact, this focused beam of solvent is omnidirectinal upon impact. Test results showed significant improvement over conventional spray techniques even at a 1 mil clearance.[9]

The use of ultrasonic cleaning on military PWAs has been disallowed for 25 years. Current military specifications do allow the cleaning of bare boards, connectors, and terminals but do now allow the cleaning of any electronic components with ultrasonics. The fear was that the vibration generated by the cavitation would apply stress to the fragile wire bonds internal to the integrated circuit. Today wire bonding technology is much more reliable and we can control the ultrasonic cleaning process more closely. Screening tests can be developed to theoretically calculate the frequency to which a particular component is sensitive. Computer simulations can then be used to calculate the internal resonant frequencies of the fragile wires and then compare this frequency to that of the primary frequency of the transducers to determine if the ultrasonic cleaning process can be used without damage to the electronic component.[10] These theoretical calculations can then be tested by evaluating the production ultrasonic cleaning process. The EMPF study that was referred to earlier will make a recommendation based upon its findings about the use of ultrasonic cleaning for military electronics.

It is critical in this day and age to continue with the development of the ultrasonic cleaning process. If we can be successful than we can use less aggressive, less hazardous solvents and accomplish more effective cleaning with the use of ultrasonics.

5.2.2. Spray Equipment

Effective defluxing requires bringing unsaturated solvent into contact with the flux residue. All equipment for alternate defluxing materials operates by either ultrasonic, spray or centrifugal mechanisms to move fresh solvent into contact with the contaminants to be removed. Spray is conducted either in air or under the surface of the liquid. Ultrasonic equipment has been discussed earlier in this chapter. Centrifugal cleaning equipment will be covered later. At this point spray equipment will be considered.

A general schematic for spray cleaning equipment for use with a terpene semi-aqueous defluxing material is given in Figure 5.5. As can be seen from this figure, spray cleaning is usually followed by a spray rinse.

The first equipment for semi-aqueous cleaning processes became available for sale in 1988. As of this writing, spray defluxing equipment is available from at least seven companies: Corpane, Detrex, Electronic Controls Design, Electrovert, Japan Field, OSL, and Vitronics.

Except for the incorporation of water rinse, the operation of spray equipment

SEMI-AQUEOUS CLEANING MACHINE SCHEMATIC

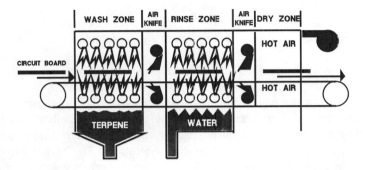

Fig. 5.5. Schematic of semi-aqueous spray equipment.

for semi-aqueous cleaning materials, as used for the alternatives described in this chapter, is essentially the same as spray equipment for use with other solvents, with the exception that there is no provision for vapor cleaning, since the alternatives are not volatile. Because the general provisions for and design of spray equipment is familiar to most readers we will not consider the process in detail. Instead, we refer the interested reader to the previously mentioned equipment vendors who can provide any necessary information on the operation of their equipment.

About seven years ago Honeywell decided to investigate the use of an inline cleaning system specifically designed for isopropyl alcohol. A very well designed system incorporating a fire arrest system with vapor monitors, a remote control panel, and even an explosion proof room was built. The system utilized a 75% isopropyl alcohol/25% deionized water solution in a combination of boiling immersion, recirculated spray and pure distillate spray. It contained a recirculating still that could recycle the entire solvent capacity in 2 hours. In performance testing it cleaned both RMA and some RA fluxes to acceptable ionic cleanliness levels. Although many dollars where invested in this system, it was not utilized for very long.[11]

In Europe, Siemens has investigated alcohol cleaning as a replacement for some CFC-113 applications. There is little likelihood that the use of alcohol cleaning will get wide acceptance in the United States because of strict fire codes.

5.2.3. Centrifugal Cleaning Equipment

Conventional cleaning equipment has traditionally used vapor, spray, immersion, or a combination of these to achieve cleaning. Recently, another process, referred to as centrifugal cleaning, has made its appearance.[12] As the name implies, this process uses centrifugal forces to cause fluid flow over the surfaces to be cleaned. Centrifugal cleaning is claimed by its proponents to be more effective than conventional methods, especially for removing contaminants from tight geometries.

In practicing centrifugal defluxing, boards are grasped by a robot head, immersed in cleaning fluid in a process chamber, and spun about an axis (usually a vertical axis) in alternating directions. The resulting motion causes the fluid to flow over the board surfaces and under any components, facilitating contaminant removal by the cleaning fluid. Angular velocity in centrifugal cleaning is analogous to spray pressure in conventional cleaning. The magnitude of the forces generated is reported to depend on the angular velocity ω, according to

$$F = M\omega^2 r$$

where

F is force,
M is density of cleaning fluid, and
r is the radial distance from the axis of revolution.

Following the wash step, the contaminated cleaning solvent is drained out of the process chamber, and rinsing is conducted in a similar fashion. A drying step completes the process.

Centrifugal cleaners offer advantages such as compatibility with multiple cleaning products, small footprint, and minimal solvent consumption. Disadvantages include the inability of some assemblies to withstand the tresses imposed by high rotational velocities. Centrifugal defluxing was pioneered by ACCEL Corp., which manufactures MICROCEL™ centrifugal cleaning equipment.

5.3.   MATERIALS

A wide variety of materials have been mentioned as alternatives to CFCs for defluxing. Most of these potential alternatives are detergent cleaners, usually with alkaline and/or saponifier action. Saponifiers and other alkaline cleaners, which are familiar to those involved in defluxing, are not the subject of this chapter. Instead, this chapter focuses on those potential alternatives which are neither halogenated solvent nor alkaline/detergent. The information about some of these cleaners is too sketchy to permit their classification. In such cases, our approach has been to include the formulations in our discussion.

In the course of preparing this information, we invited manufacturers of all of the alternative materials of which we had knowledge to submit information

Table 5.1. Material Properties.

	ACT 100	AXAREL 38	BIOACT EC-7	BY-PAS	M-PYROL	MARCLEAN
Type	NA	Semi-aqueous	Semi-aqueous	Alkaline	Semi-aqueous	Semi aqueous
Composition	NA	Prop.* hydrocarbon	Terpene/NA surfactant mixture		NMP	Prop.*
Boiling Point °F	230	400	340	208	396	428
Flashpoint, °F	none	159 TCC	160 COC	none	204 COC	240 TCC
Density, g/cc at 25°C	0.99	0.85	0.84	1.08	1.03	NA
Vapor pressure, mm Hg at 25°C	<2	0.2	1.6	NA	<1	0.06
Viscosity, cP at 25°C	NA	1,4	0.8	NA	1.7	17
pH, 5% in H$_2$O 10%	NA	5–6	5–6	NA	7.0	5–6
Freezing point, °F	NA	NA	≤40	NA	− 12	44
Surface tension, dynes/cm	NA	28	33	NA	NA	NA
Odor	low	mild	citrus	mild	mild	low
BOD$_5$	NA	423	1500	NA	NA	NA
COD g/l	NA	1600	2500	NA	NA	NA

NA = Information not available.
*Proprietary material
Source: Product Manufacturers

relative to the use of their products for defluxing. While a significant number of products have been touted as potential CFC alternatives, only the six products described below met the criteria outlined above. This was the state of the art in late 1989.

The source of the list of materials considered was the list of potential CFC alternatives for use as defluxers as developed by the IPC solvent alternatives group. While this standard is somewhat arbitrary, we think it is a reasonable one in view of the eminent studies of the IPC in the electronics industry, and the seemingly reasonable presumption that those manufacturers who offer serious candidate materials would probably participate in the IPC activities.

We have summarized the information received in this data collection process in Table 5.1. Six alternative materials met these tests and are included in the table.

The nature of the task made it necessary for us to accept as fact the data provided by the product manufacturers. Thus, these data are not necessarily endorsed by the authors, though we have attempted to verify facts wherever possible. One of the problems typical of emerging technologies is reflected here— that the volume of information available for some of the alternatives is vastly more than is available for others. While this may appear to result in an incompletely balanced presentation we believe it is preferable to provide the reader with as much detail as is available even if that means that some materials are more thoroughly covered than others. In the sections that follow, we have specifically reviewed the possibility of using the following products as CFC defluxing alternatives:

1. Advanced Chemical Technologies' ACT-100
2. Du Pont's AXAREL 38
3. Petroferm's Bioact ^{R} EC-7™ (marketed by Alpha Metals)
4. By-Pas of Toledo's By*PAS
5. GAF's M-Pyrol
6. Martin Marietta's MarClean

5.3.1.   Performance

Performance is a word which expresses an absolutely critical concept, but which encompasses a broad spectrum of topics. We will consider here some of the more important of these topics, including (1) cleanliness testing, (2) contaminant solubility in the cleaning solvent, (3) surface tension and contact angle of the cleaning medium, (4) compatibility of the cleaning composition with the electronic assemblies' materials of construction, and (5) process control necessary or possible for a given cleaning product.

Cleanliness Testing.   Cleanliness testing is used to assess how well a material performed in a cleaning process. In most cases the test measures a coupon after cleaning to determine or characterize what contaminants have not yet been removed. In the past, practical cleanliness testing was limited to ionic cleanliness testing. But in the last few years several other test methods have been used to characterize the remaining residues. Other techniques include residual rosin, surface insulation resistance, (SIR), and high performance liquid chromotography (HPLC).

Ionic cleanliness is by far the most common technique used and is easy to perform with commercially available equipment. A general description of an ionic cleanliness test is given here. Typically a test is performed using a coupon prepared using the same cleaning material, equipment and process as is used in the manufacturing environment. The coupon is then immersed in an essentially ion-free (greater than 20 Megohm-cm resistivity) 75% isopropyl alcohol/25% deionized water solution and the resistivity change is monitored. The results are then compared to a standard sodium chloride solution resistivity change. The results are expressed as micrograms of contamination per square inch of coupon area as compared to a sodium chloride equivalent.

Contaminant Solubility. The cleaning performance of a material can be evaluated both theoretically and practically. Published values exist for many of the more common properties of materials. Specifically, the Kauri-butanol (Kb) value and the Hansen solubility parameter are two indices that are used to compare solvent power. The Kauri-butanol value relates to the solvency of the material for Kauri rosin. The higher the value the better the solvent.

The application of Kauri-butanol values to defluxing is somewhat limited because of differences between the Kauri rosin and gum rosin.[13,14] The Hansen or Hildebrand solubility parameters are more applicable to the defluxing process. The concept used with solubility parameters is that the cleaning material solubility parameter should closely match the solubility parameter of the contaminant being removed. In this case a higher value is not necessarily better. The Hidebrand solubility parameter for gum rosin varies from about 19 to about 21, depending on the particular circumstances.

These theoretical estimates are frequently useful as a guide in selecting candidate materials for testing as defluxing solvents. The question is settled in practice by actually measuring the solubility of flux residue in the solvent in question. Abietic acid is often used as a model compound for flux residue. Good defluxing solvents tend to have high affinity for flux residue, though exceptions can occur.

Surface Tension and Contact Angle. In addition to the solubility of flux residue in a proposed defluxing solvent, another important factor is the fluid's ability to penetrate into/under closely spaced components to remove flux residues from tight geometries. Often, this question assumes the form, "What is the surface tension of . . . (the product in question)."

Several common cleaning fluids with low surface tension penetrate well into small spaces under closed-mounted components so, unfortunately, the mistaken perception has arisen that good penetration is due to low surface tension.[13-17] The fact is that good penetration on the part of low surface tension liquids such as CFC-113, methyl chloroform, methylene chloride, and the like is in spite of, not because of, their low surface tension. Solvents tend to wet the soiled surfaces under surface mount technology (SMT) components better than water does; this, and not their lower surface tension is the principal reason solvents clean tight geometries better than water does. High surface tension actually aids penetration, all else being equal.

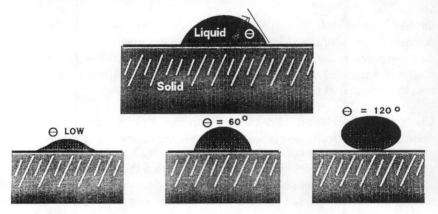

Θ IS THE ANGLE FORMED BETWEEN THE SURFACE OF THE SOLID
AND THE TANGENT TO THE LIQUID AT THE SOLID/LIQUID/GAS
INTERFACE. AS MEASURED THROUGH THE LIQUID.

Fig. 5.6. Definition and illustration of contact angle.

A more fundamental determinant of penetration ability is contact angle. The contact angle of a liquid on a solid is defined as the angle formed between the surface of the solid and the tangent to the liquid at the solid/liquid/gas interface, measured through the liquid. This is illustrated schematically in Figure 5.6. Contact angle is one of the most significant factors determining the performance of a cleaning product in removing contaminants from close-tolerance spaces.

The importance of contact angle is readily illustrated by reference to Figure 5.7. Here, three cases are shown: high, medium, and low contact angle. It is immediately apparent that all else being equal penetration is easier and more likely to occur in the case of low contact angle. In contrast, surface tension is relatively unimportant.

There are several references available for those who might wish to more fully explore the relationship between cleaning and penetration ability.[13,14,18–20] Especially recommended is the work of Ostrovsky,[18] an elegant, comprehensive treatise concerning the theoretical factors affecting post-solder cleaning processes for printed wiring assemblies. This work demonstrates unequivocally that surface tension is not the dominant factor determining penetration and cleaning in close-tolerance spaces.

Materials Compatibility. Many cleaning product companies supply compatibility data as part of a materials technical data sheet. This information is helpful especially where plastics and elastomers are concerned, but more specific information is often necessary. Other critical compatibility tests are also necessary

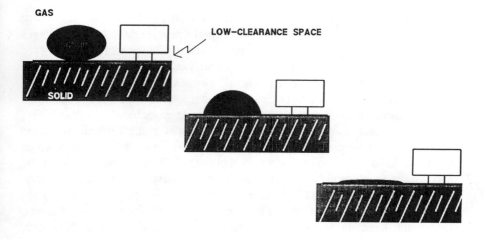

Fig. 5.7. Illustration of the effect of contact angle on liquid penetration.

to determine if other process chemicals are compatible. Laminate material such as G-10, FR-4, polyimide, or Kapton should be evaluated for compatibility with the cleaning solvent and process. Especially susceptible to incompatibility problems are electronic part markings and permanent solder masks.

Process Control.   A cleaning material is only as good as the process it is used in. If the process is complex and needs constant monitoring, the margin for error increases. The optimal situation is one where a material is loaded into the equipment and requires no attention until the process is completed or the material is depleted. Automatic filling equipment solves the rather easy problem of material replenishment. In some materials chemical analysis is required to determine when a material should be replenished or when saturation point has been reached. In the selection of a particular cleaning material it is important to evaluate the process control requirements. The material manufacturer may or may not have specific process variables such as temperature, cycle times, spray pressures and flow rates listed in the product data sheet. Usually the manufacturer will provide some guidelines for use, but it is up to the end user to optimize the process to his or her specific needs. Many new alternatives do not have established processes yet so the burden for this testing remains with the end user.

5.3.2.   Environmental Acceptability and Safety Considerations

Environmental Acceptability.   The major reason for the widespread interest in alternative defluxers is the detrimental environmental effects of halogenated solvents, especially stratospheric ozone depletion due to CFCs. Some alternative manufacturers claim that their products offer performance and/or cost advantages relative to CFCs, but environmental concerns are still the primary driving force causing most assemblers to attempt to identify alternative defluxing agents, sometimes only to trade one environmental problem for another one.

In determining the general environmental acceptability of a CFC substitute candidate, some of the factors that should be considered include biodegradability, recyclability, ease of disposal, energy consequences and effects of employee exposure. Each potential alternative will have its own set of characteristics which determine its overall environmental acceptability. Tradeoffs must usually be made, since all or virtually all chemical substances have some undesirable attributes. For example, CFC-113 (and its commonly used azeotropes) is very readily recycled, the energy cost of using it is acceptable, and its toxicity is low. Further, storage and handling of CFC-113 products is safe and convenient. However, CFCs are extremely resistant to degradation. This inertness is a primary reason that they survive unaltered in the environment for very long times. The CFC molecules eventually migrate to the stratosphere where they slowly decompose to release chlorine atoms, which then react with ozone molecules in a very complex process. So destructive is this process that one molecule of CFC-113 may destroy 300,000 molecules of ozone. To put this in perspective, consider that the average amount of CFC-113 used to clean one square meter of circuit board (3 kilograms, based on U.S. figures for a recent year) contains enough chlorine to destroy one hundred thousand cubic meters of ozone ($100,000$ m^3), measured at standard temperature and pressure. This balance of properties is an example of why it is necessary to consider the full range of attributes of a candidate material.

In the particular case of CFCs, it has been determined that the consequences of CFC-caused stratospheric zone depletion outweigh all of their advantage. So, while the alternatives may not be environmentally perfect, some are available which appear to offer environmental advantage over the CFCs.

Before making a commitment to use a CFC alternative, an electronics assembler should work with the chemical material supplier to determine that the candidate product can be used in his operations without unacceptable environmental consequences.

Safety Considerations. There are two aspects of concerns about employee safety with respect to use of defluxing compositions: toxicity and physical safety. Fire is the only plausible physical hazard associated with any of the materials addressed in this chapter. That concern can be addressed for all these materials collectively, since there are no substantial differences among them in terms of their combustibility under proposed conditions of use. We review and summarize the situation below.

On the other hand, with respect to toxicity, each material must be considered individually, because of their distinct chemistries. We have attempted to provide the reader with guidelines (from the Code of Federal Regulations) for assessing the toxicological risks resulting from exposure to chemical substances, given certain minimum information. We are pleased to note that, insofar as we can determine from the available literature, the substances proposed to date as semi-aqueous cleaners do not appear to pose any significant hazard.

Flammability.   To the best of our knowledge, all of the materials proposed as semi-aqueous cleaners are capable of being ignited under some conditions. With the exception of isopropyl alcohol, none of the cleaners mentioned to date are classified as flammable substances. Flammability is a term which is frequently misunderstood. Materials which have flash points (closed cup) below 100° F are classified as flammable.

Flash point is the temperature at which sufficient vapors exist over a liquid that if a flame is brought into the vapor space immediately above the liquid, it will ignite the vapors. There are two ways of measuring flash point: closed cup and open cup. As the name implies, these methods are distinguished by the way in which the experimental apparatus is arranged.

For determining closed cup flash points, common examples of which include Tag and Pensky-Martens methods, vapors are forced to accumulate in a confined space as the temperature is gradually raised. A gas pilot flame is periodically brought into contact with the accumulated vapors. This is accomplished via a small port which is kept closed except for a brief time at the moment of testing. At the flash point temperature, when the port is opened and the flame is brought near, a flash fire results which burns off the accumulated vapors and then goes out.

In the open cup flash point test, the vapors are not confined. Instead, a gas pilot flame is kept burning continuously a specified distance above the surface of the test material contained in a standard cup which is gradually heated. In both closed cup and open cup testing, the flash point is defined as the lowest temperature at which a flash fire occurs.

As noted above, substances are classified as flammable if they have closed cup flash points below 100° F. They are designated as combustible if their flash point is between 100°F and 200°F. There is no further classification for those materials having flash points above 200°F. Most of the compositions proposed as semi-aqueous cleaners to be used as CFC alternatives are combustible substances.

In cleaning machines which operate by means of spraying in air, flash point has little to do with fire safety. This is true provided that the cleaning process is carried out below the flash point. Below the flash point temperature, any spark or flame which might occur could cause a fire. Above the flash point, a fire is not possible except under conditions where mist may occur.

Mist combustibility is virtually independent of the flash point of the liquid forming the mist. For example, Eichhorn found that flame propagation characteristics of Stoddard solvent (closed cup flash point 110°F) were indistiguishable from those of peanut oil (open cup flash point 615°F).[21] Thus, mist can be combustible even at temperatures far below the flash point of the bulk liquid. The somewhat simplified reason for this is that an individual mist droplet can burn if it is raised to a temperature above its flash point. An ignition source might ignite one such droplet which, upon burning, may heat nearby droplets and cause the flame to spread to them, where the process is repeated.

Mists are much harder to ignite than vapors, but a suficiently intense ignition source can cause mists to burn. In the case of mists, the principal factors determining whether combustibility is possible are particle size and density of particles (number per unit volume). In order for mist to be combustible, the particle size has to be sufficiently small (a rule of thumb is that particles must be smaller than about 1500 microns in diameter) and the droplets must exceed a minimum density level, in terms of particles per unit volume.[22] That is, the population of mist droplets in a given space must be sufficiently high that the flame can propagate from one particle to the next.

It is difficult to make any general quantitative statement concerning the minimum mist density required to allow flame propagation, but experimentally it is observed that a rather dense mist is required and that with a sufficiently high density, there is a theoretical possibility of a very rapid fire or even an explosion. The minimum mist density required for explosivity is well worked out. This value is about 48 grams of mist plus accompanying vapor per cubic meter of air (at 0°C and atmospheric pressure).[22–24] A mist of this density takes the form of an extremely thick fog; in it the visibility is so low that a 100 watt light bulb can be seen for a distance of only a few inches.[23,25] It is extremely rare that such conditions occur.

Another aspect of ignitability is the question of the upper and lower explosive limits. Essentially, these two values give the concentration range between which vapors of the substance in question can be ignited. The lower explosive limit represents the minimum amount of the substance's vapor necessary to sustain combustion. The upper explosive limit represents the minimum amount of oxygen required or, in other words the maximum concentration of vapors which can be present and yet which allows the presence of enough oxygen to sustain combustion.

In producing equipment for using the alternative products described in this chapter, attention to fire prevention details is necessary. Two approaches are available. One of these is to design the cleaning machine in such fashion that it operates below the flash point of the cleaning composition, and such that any mist formed does not constitute a hazard. The other method of providing fire protection is to maintain an atmosphere in the wash section of the machine which is too low in oxygen to permit combustion. This is done by blanketing the area with inert gas, usually nitrogen. As of this writing, all of the commercially available equipment for CFC alternative materials utilizes one or the other of these fire prevention methods.

Employee Exposure. The semi-aqueous cleaners proposed to date and about which information is available are not significantly hazardous in terms of employee exposure. The Code of Federal Regulation (CFR), Section 1910.1200, Appendix A sets forth definitions of various health hazards. Appendix B of the same section outlines the principles and procedures of hazard assessment.

Chemicals which meet any of the following definitions, as determined by the criteria set forth in Appendix B of CFR 1910.1200 are health hazards:

1. Carcinogen. A chemical is considered to be a carcinogen if:
   a. It has been evaluated by the International Agency for Research on Cancer (IARC), and found to be a carcinogen or potential carcinogen; or
   b. It is listed as a carcinogen or potential carcinogen in the *Annual Report on Carcinogens* published by the National Toxicology Program (NTP) (latest edition); or,
   c. It is regulated by OSHA as a carcinogen.
2. Corrosive. A chemical that causes visible destruction of, or irreversible alterations in, living tissue by chemical action at the site of contact. For example, a chemical is considered to be corrosive if, when tested on the intact skin of albino rabbits by the method described by the U.S. Department of Transportation in Appendix A to 49 CFR Part 173, it destroys or changes irreversibly the structure of the tissue at the site of contact following an exposure period of four hours. This term shall not refer to action on inanimate surfaces.
3. Highly Toxic. A chemical falling within any of the following categories:
   a. A chemical that has a median lethal dose (LD_{50}) of 50 milligrams or less per kilogram of body weight when administered orally to albino rats weighing between 200 and 300 grams each.
   b. A chemical that has a median lethal dose (LD_{50}) of 200 milligrams or less per kilogram of body weight when administered by continuous contact for 24 hours (or less if death occurs within 24 hours) with the bare skin of albino rabbits weighing between two and three kilograms each.
   c. A chemical that has a medial lethal concentration (LC_{50}) in air of 200 parts per million by volume or less of gas or vapor, or 2 milligrams per liter or less of mist, fume, or dust, when administered by continuous inhalation for one hour (or less if death occurs within 1 hour) to albino rats weighing between 200 and 300 grams each.

4. Irritant. A chemical which is not corrosive but which causes a reversible inflammatory effect on living tissue by chemical action at the site of contact. A chemical is a skin irritant if, when tested on the intact skin of albino rabbits by the methods of 16 CFR 1500.41 for four hours exposure or by other appropriate techniques, it results in an empirical score of five or more. A chemical is an eye irritant if so determined under the procedure listed in 16 CFR 1500.42 or other appropriate techniques.

5. Sensitizer. A chemical that causes a substantial proportion of exposed people or animals to develop an allergic reaction in normal tissue after repeated exposure to the chemical.

6. Toxic. A chemical falling within any of the following categories:
   a. A chemical that has a median lethal dose (LD_{50}) of more than 50 milligrams per kilogram but not more than 500 milligrams per kilogram of body weight when administered orally to albino rats weighing between 200 and 300 grams each.
   b. A chemical that has a median lethal dose (LD_{50}) of more than 200 milligrams per kilogram but not more than 1,000 milligrams per kilogram of body weight when administered by continuous contact for 24 hours (or less if death occurs within 24 hours) with the bare skin of albino rabbits weighing between two and three kilograms each.
   c. A chemical that has a median lethal concentration (LC_{50}) in air of more than 200 parts per million but not more than 2,000 parts per million by volume of gas or vapor, or more than two milligrams per liter but not more than 20 milligrams per liter of mist, fume, or dust, when administered by continuous inhalation for one hour (or less if death occurs within one hour) to albino rats weighing between 200 and 300 grams each.

7. Target Organ Effects. A number of specific effects are listed.

One further aspect of employee exposure which may be worth addressing concerns airborne vapors of semi-aqueous, low-volatility cleaners. There are no PEL (permitted exposure limit) or TLV (threshold limit values) established for most of the alternatives described in this chapter. However, a typical suggested PEL for an alternative product is 400 ppm. By definition, a TLV or PEL value sets a standard concentration below which no harmful effects are expected. The ratio of PEL to vapor pressure gives a measure of the relative probability that exposure may exceed the recommended level under similar physical conditions such as temperature, ventilation, etc. Alternatively, this ratio can be looked upon as a simple measure of the degree of vapor control necessary to maintain exposure below the PEL. The higher the ratio, the greater is the safety margin. The vapor pressure of CFC-113 is approximately 180 torr at room temperature. Its PEL of

1000 ppm gives a ratio of 1000/280 or 3.6 for CFC-113, a substance generally regarded as very safe in use. Most of the products discussed in this chapter have vapor pressure between approximately 1 and 2 torr. Taking a PEL of 400 ppm gives a PEL to vapor pressure ratio of 400/2 or 200. Clearly, the probability of having vapor from such a material at a concentration exceeding the recommended PEL is extremely remote. In fact, assuming that the PEL's are appropriate exposure limits, overexposure would then be found to be 55 times less likely than CFC-113 overexposure (given by 200/3.6 = 55).

Relative to methyl chloroform, these products are safer by an equally overwhelming margin. For methyl chloroform, the TLV is 350 ppm and the vapor pressure at room temperature is 100 torr, so the PEL to vapor pressure ratio is 350/100 to 3.5. Hence, overexposure to semi-aqueous products is almost 60 times less likely than overexposure to methyl chloroform.

Another concern in switching from CFC-113 cleaners to semi-aqueous processes is the possible VOC emissions. The basis for this concern is that CFCs (and most chlorinated solvent compositions) are not classified as VOCs, while essentially everything else is. In fact, however, much CFC-113 is used in the form of azeotropes which contain some substances which are themselves VOCs. Switching from CFC-113/methanol azeotrope (6% methanol) to a low-volatility cleaner almost always reduces the mass of VOC emissions, sometimes quite substantially.

As a first approximation, all CFC products are eventually discharged to the atmosphere. In contrast only a very small proportion of a semi-aqueous product evaporates. As noted earlier, a typical CFC use rate in printed wiring assembly defluxing is 2 kilograms per square meter. It was also noted that, with substantial effort, it may be possible to reduce CFC consumption to about 1 kg per square meter of broad area cleaned and, with highly capital-intensive modifications such as carbon absorption vapor recovery units, consumption may be reduced by 0.5 kg/m^2. Reducing use below 0.5 kg/m^2 is not likely. Thus, per square meter of board surface cleaned with "non-VOC" CFC 113/methanol, actual VOC emissions are typically more than 100 grams (2000 g $\times$ 0.06 = 120 g). VOC emissions from a typical semi-aqueous cleaning process are much lower. Total semi-aqueous cleaner consumption is usually in the range of 5–10 grams per square foot of board cleaned, or approximately 50–100 grams per square meter. Most of this cleaner, along with the dissolved contaminants, is rinsed off the board and processed with the waste water. Exact numbers are not yet available, but the best estimate suggests that not more than about 5 grams of VOC would be emitted per square meter cleaned. To put this in perspective, it means that switching from CFC-113 azeotrope to semi-aqueous cleaning would reduce VOC emissions by a factor of about 5, even in the conservative case. Relative to current practice, in which 2 kg per square meter of CFC is consumed, switching from CFC to semi-aqueous cleaning would reduce VOC emissions by a factor of more than 20. Most people find this surprising, because the facts have not been well understood.

5.3.3. Economics

In order to have a meaningful discussion of economics of alternatives, it is first necessary to understand the economics of CFC use. Determining the cost of chemicals used in defluxing circuit boards should be rather simple, but in fact it is the subject of quite a lot of discussion. Many assemblers claim that the amount of CFC used for cleaning boards is much less than the CFC manufacturers report; this is addressed below in more detail.

Equipment Costs.   Aside from chemical consumption and operating costs, the other major factor in the economics of alternative defluxing is the equipment required. Equipment for use with the alternatives described in this chapter is substantially different from the equipment used for CFCs. Despite that, it appears as though equipment costs can be expected to be roughly comparable, provided that it is not necessary to put carbon adsorption or other rigorous controls on the CFC equipment. If these controls are necessary, semi-aqueous cleaning equipment may be substantially less expensive, since stringent emission control equipment for CFCs is itself quite expensive. The best estimate at this point seems to be that equipment costs for alternatives will be no greater than for CFCs or other halogenated solvents.

Simple, batch vapor degreasers will soon be things of the past. While these units are low in cost and easy to operate, they are extremely wasteful of solvent. The need for solvent conservation, coupled with the rapidly escalating cost of halogenated solvents will almost certainly mean the demise of unsophisticated batch vapor degreasers.

Instead, the new generation of batch vapor degreasers will include many solvent-conserving features. As a result, this equipment will cost much more than batch cleaners previously cost. Batch semi-aqueous cleaner costs are likely to be competitive with costs of sophisticated batch vapor degreasers.

In the case of inline equipment, the situation is similar. Modern, high-capacity solvent cleaning equipment costs approximately the same as semi-aqueous cleaning equipment of comparable capacity.

Use Rate and Economics of CFC Defluxing.   One way to get an industrywide estimate of the cost of using CFC to clean a unit area of circuit board is to take the volume of CFC product sold for defluxing, along with its total cost (cost per gallon times total gallons), and divide by the total surface area of boards cleaned. This approach gives a number that is correct on a macroscopic basis, whatever the merits of any individual situation. Some people quarrel with this because the resulting cost figures are rather high.

Industry figures for a recent year show that, in the United States, about 70 million pounds (30 million kilograms) of CFC-113 products were used to deflux about 100 million square feet (10 million square meters) of printed wiring boards. (These numbers are rounded to one significant figure.) At an average price of $1.00 per pound (which is probably substantially less than the current average price of CFC products), this leads to a defluxing cleaner cost of about $ 0.66 per square foot ($ 6.60 per square meter) of board cleaned.

The accuracy of this cost number depends only on the accuracy of the figures for the amount of CFC sold into this application and on those for the board area cleaned. Many assemblers are surprised to learn that defluxing with CFC costs so much. It is not unusual for contract assemblers to include a cleaning charge of $ 0.25 per square foot ($2.50 per square meter) in their service agreemments. On the average, this looks like quite a bargain.

Another way to estimate the cost of cleaning with CFC is to take an account of the amount of CFC product required to clean a unit area of board in a particulat cleaning machine. Numbers have recently been reported for three cases.[26,27] The first case, referred to here as the typical case, is said to represent a more or less average condition in which there is neither excessive waste nor significant attempt to control CFC emissions. In that instance, the use rate of CFC product for cleaning is approximately two kilograms per square meter. The cost of the defluxing chemical would come to about $ 0.44 per square foot ($4.40 per square meter). This use rate can be considered in good agreement with the 3 kg per square meter figure deduced earlier, since some of the CFC which is sold for defluxing may in fact be used for other jobs, once on the plant floor.

The imposition of good housekeeping practices and careful equipment operating procedures can sometimes reduce the use rate by half, to 1 kilogram per square meter. Use of the best available controls, such as carbon adsorption, in connection with very stringent process controls can sometimes reduce the CFC consumption rate to 500 grams per square meter of board surface cleaned. In these cases, the cost of defluxing cleaner would be approximately $0.22 per square foot ($2.20 per square meter) and $0.11 per square foot ($1.10 per square meter), respectively.

Use Rates of Alternatives.   All of the alternative products considered in this chapter are relatively nonvolatile. For that reason evaporation, the principal loss mechanism in the CFC case, is not a significant factor for these alternatives. Instead, the principal loss mechanism with nonvolatile alternatives is dragout. Evaporation losses are minor and can be estimated at 10% or less (as a maximum) of the total consumption. While extensive data are not yet available (and preliminary data are available only in the case of terpenes), it is reasonably clear that consumption rates can be expected to be roughly comparable for all non-volatile semi-aqueous alternatives.

In a semi-aqueous process, the two major factors controlling consumption are the viscosity of the cleaning material and the manner in which the optional air knife is used to remove excess material from the boards as they exit the wash section (see Figure 5.5). Experience to date indicated that the use rate will be on the order of 5 grams of wash bath liquid per square foot of surface, but may range as high as 10 grams per square foot in some circumstances. As noted earlier, of that amount about 90% (at a minimum) will be taken into the rinse water, and 10% or less will evaporate.

Economics of Alternative Defluxing.   From the unit volume costs of the alternatives, the consumption rate allows for estimation of costs relative to the cost of cleaning with CFCs. Firm prices for most CFC alternatives have not yet been established. Nonetheless, the arithmetic of the situation makes it clear that cleaning with alternatives described in this chapter can be expected to be very competitive with, and in some cases substantially less costly than, cleaning with CFCs. This may be surprising in view of the commonly discussed possibility that alternative halogenated solvents will cost significantly more than currently available CFCs, and use rates are likely to be similar as for the presently used CFCs. If that is true, the alternative materials discussed in this chapter appear especially attractive.

Taking a specific example, consider an efficiently operated defluxer using CFC-113 (or an alcohol azeotrope). From the information presented earlier, the CFC consumption can be estimated at 1 kilogram per square meter, and the consequent chemical cost is approximately $2.20 per square meter ($0.22 per square foot). For semi-aqueous cleaner, at a use rate of 100 grams per square meter (10 grams per square foot), defluxing chemical cost is less for the semi-aqueous process than for CFC cleaning as long as the price of semi-aqueous product is not more than about $70 per gallon (they are likely to be far less expensive than that). This fact serves to point out the significant cost advantage held by nonvolatile cleaners over CFCs, even if the semi-aqueous products cost somewhat more per gallon.

5.3.4.   Specific Material Descriptions

We will now consider in more detail six specific products which have been proposed for use in defluxing printed circuit assemblies. Some are relatively well known; some are virtually unknown in the assembly industry. Virtually all of the information in this section is based on written materials supplied to the authors by the companies marketing the respective products. We have attempted to verify facts wherever possible but, for the most part, we have little first-hand knowledge about them.

1. Advanced Chemical Technologies ACT-100. ACT-100 was developed from a positive photo resist stripper. This product was evaluated for biodegradability and according to studies performed by Lehigh University and is classified as biodegradable. This material is incompatible with some elastomers and softens Teflon. It does not attack part markings metals and substrates. In order to ensure long life of equipment used with this product stainless steel, teflon and butyl rubber are recommended.

ACT-100 is a mild skin irritant. Skin should be flushed with water for 15 minutes if direct skin contact with the material is made. It is also classified as a severe eye irritant. This material can be recycled, however in small volumes it may not be cost effective to recycle. The manufacturer is developing a recycling program for large users of the material. There are several formulations of ACT-100 available for flux removal. A new formulation is soon to be available specifically formulated for defluxing surface mounted assemblies. Spray cleaning and/or the use of ultrasonics yields successful results with this material. The cleaning process that is recommended by the manufacturer is a three step process. The first step is immersion of the workpiece in a heated (50–60°C) solution of ACT-100 with the use of spray or ultrasonics. The second step is an agitated ambient water rinse. The last step is either a hot air dry or an alcohol rinse to dry.

2. Du Pont's AXAREL 38™. AXAREL 38 is a proprietary hydrocarbon, semi-aqueous cleaning material with polar and nonpolar constituents formulated to remove white residue. Tests were performed using AXAREL 38 with both rosin activated (RA) flux and rosin mildly activated (RMA) flux. Figure 5.8 shows performance equal or better to that of an existing CFC-113/methanol azeotrope. Note that AXAREL 38 is referred to in this figure by its developmental name 9438. The results in this figure are expressed as a per cent of military specification requirement which is 21 micrograms NaCl per square inch. Surface insulation resistance (SIR) showed satisfactorily high results under high humidity and temperature conditions.

AXAREL 38 has a moderate flash point (158°F) and is classified as an NFPA Class III A combustible liquid. Fine mists can be ignited. The simplest precaution with a mist is to use an application method that avoids misting such as spray-under immersion, spin-under immersion, or ultrasonics. Use an inert atmosphere if direct spray is required. As with many semi-aqueous processes the best process for this material involves the use of a spray/immersion step followed by a water rinse step. This material is best used at room temperature or slightly above and at 100% concentration. Batch processing requires a four minute immersion in

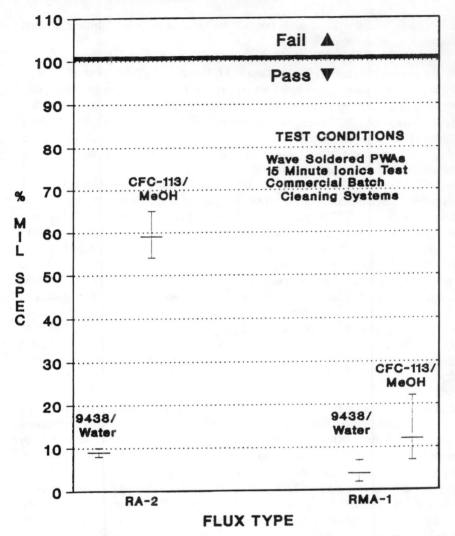

Fig. 5.8. Ionic contamination removal using Axarel 38. (Reprinted with permission from K.T. Dishart and M. C. Wolff, E. I. Du Pont de Nemours. "Advantages and Process Options of Hydrocarbon based Formulations in Semi-Aqueous Cleaning". Presented at NEPCON West Anaheim CA 2 2690)

the material followed by a four minute water rinse. Average spray pressures of 30–50 psig can be used but this can be lowered with increases in flow rate.

As with many hydrocarbons material compatibility of this material is a concern. Tests were performed on various materials and after a 24 hour immersion at 50°C some plastics and elastomers were affected. This material is considered biodegradable and should be acceptable to most wastewater treatment plants. This material can be easily separated using a water/hydrocarbon separator. The hydrocarbon and soils can be incinerated as fuel. The aqueous effluent can be treated in a closed loop system. This material is classified as a volatile organic compound (VOC) but it has low volatility and therefore emissions are negligible under proper use conditions.

3. Petroferm's BIOACT EC-7®. Terpene/surfactant mixtures have been used for cleaning electronic assemblies for several years.[28,29] BIOACT EC-7 is a commercially available terpene-based defluxer. Prior to the recent wave of concern about CFCs and ozone depletion, EC-7 was sold on the basis of superior defluxing performance relative to CFC-113/methanol and its easier waste treatment characteristics compared to aqueous detergents (saponifiers). BIOACT EC-7 is composed of approximately 90% terpene hydrocarbons (derived from citrus fruit processing) and approximately 10% nonionic surfactants. The product is made by Petroferm Inc., which holds patent rights for the use of terpenes in cleaning electronics assemblies. Terpenes are naturally derived, biodegradable solvents. They are chemically similar to rosin, the major flux component, and dissolve flux residue extremely well. They have a long history of contact with people; their low toxicity is well established. Many terpenes are listed in the Code of Federal Regulations as GRAS (generally recognized as safe). The only significant point of concern in the use of terpenes as defluxing solvents is their combustibility. The fire safety issue has been satisfactorily addressed by cleaning equipment manufacturers. The material compatibility is similar to that of CFC except that components must also be able to withstand water immersion.

Terpenes are capable of providing satisfactory cleaning for each of the commonly available flux types: RA, RMA, SA, and OA. This is illustrated in Figure 5.9, where cleaning performance is assessed for BIOACT EC-7. Data are presented comparing EC-7 with CFC-113/methanol for cleaning RMA flux and comparing EC-7 with water cleaning for an OA (water soluble) flux. Further comparative data, including production cleaning results are available in the literature.[30-39] The conclusions from each of these studies are similar to those in Figure 5.9—terpenes clean at least as well as halogenated solvents.

Equipment for use with terpenes is available from several leading cleaning equipment suppliers. Batch units are offered by Accel and ECD. Continuous, conveyorized, in-line cleaners are built by Corpane, Detrex, Electrovert, Japan

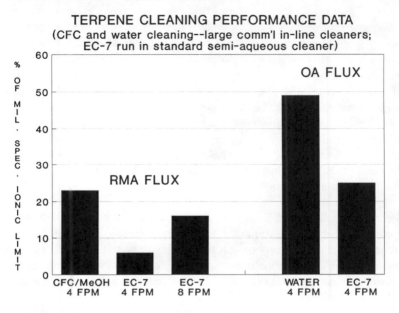

Fig. 5.9.   BIOACT EC-7 cleaning performance data.

Field, OSL and Vitronics. As of this writing terpenes are being used for electronics cleaning by more than 100 customers.

In May, 1990 the EPA/DOD/IPC AdHoc Solvents Working Group completed testing of BIOACT EC-7 and indicated that the material met the criteria established by the Group for approval as an acceptable CFC alternative for removal of rosin solderpaste and rosin flux from electronic subassemblies. Detailed reports are available from the material manufacturer and the IPC.

Recently, Petroferm/Alpha Metals have introduced a second member of the EC-7 family, BIOACT EC-7R™. It has the same performance characteristics as the original EC-7, but offers other properties which may be of advantage to some users. EC-7R is designed to separate quickly and completely from the rinse water. It is used in standard semi-aqueous cleaning equipment. A process schematic is given in Figure 5.5.

EC-7R is designed to address the principal disadvantage of semi-aqueous cleaners (and water cleaning) relative to solvent (e.g., CFC) cleaners, namely, water disposal. With EC-7R, using only a gravity separator, the terpene plus contaminants phase can be easily separated from the aqueous phase. The separation occurs within a few minutes at any temperature from 0°C (32°F) to 100°C (212°F). The resulting water phase is clean enough that it can be discharged to the sewer without further pretreatment; typical COD levels are at or below 100 parts per million. Alternatively, the water can be cycled through an additional

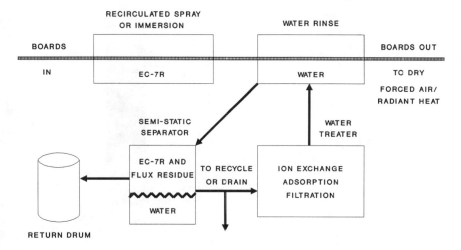

Fig. 5.10.   Closed Loop Semi-Aqueous Cleaning Process.

treater and returned to the process, thereby eliminating any water effluent at all (see Figure 5.10).

4. By-Pass of Toledo.   By-Pas is a multi-purpose cleaning material that has been introduced to the defluxing market. This material is a biodegradable emulsifying agent. It is non-toxic, non-flammable material and can be used in a variety of concentrations depending upon the intended cleaning application. According to the manufacturer, used material can be disposed into the main drain. The only storage precaution is to prevent from freezing. As with other semi-aqueous materials, after immersion in By-Pass rinsing with water is required.

5. GAF's M-Pyrol.   M-Pyrol is also known as N-methyl-2-pyrrolidone or NMP and has been widely used in many industries as a chemical reaction medium and as a formulating agent. Although M-Pyrol is an aggressive solvent at 100% concentration, lesser concentrations have been very effective in ionic cleaning testing. GAF is the prime manufacturer although this material is available from many other companies as well. In order to evaluate the biodegradability of M-Pyrol tests were conducted using the technique of WARBURG respirometry. It was determined that the biochemical oxygen demand (BOD) values were acceptable. The overall toxicity of M-Pyrol is relatively low although it is not to be take internally. It is a mild irritant to the skin and should be rinsed off immediately if skin contact has been made. It is also a moderate eye irritant.

The storage of this material is of some concern because of its hygroscopic

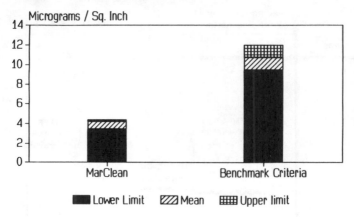

Fig. 5.11.   MarClean cleaning performance data.

properties. The shipping and storage containers of the material should be stored in a dry area. When using a drum of the material care should be taken with the partially full drum to prevent entry of moisture from condensation.

Manufacturers claim that through the use of vacuum distillation 90% of the material can be recovered. Test were performed on various O-ring gaskets and hoses and some materials were found to be incompatible (including Viton A). Users should check compatibility of individual materials. M-Pyrol has a relatively high flash point (199°F closed cup). Because of the materials low volatility evaporation losses are almost eliminated. The cost of the raw material is approximately $1.50 per pound. If blends are used then total material cost would be lowered. Various processes are successful using M-Pyrol straight immersion, spray wash or ultrasonics or a combination of the processes listed above. This material tends to darken upon age. The darkening does not affect product performance and is thought to be caused by an aldehyde formation in the parts per billion concentration range.

Martin Marietta's MarClean™.   MarClean is a semi-aqueous cleaning material designed to clean surface mount and through-hole printed wiring assemblies. In performance testing by the manufacturer this material cleaned better than the IPC Benchmark CFC-113/methanol azeotrope in both ionic cleanliness and residual rosin testing. See Figure 5.11 for a comparison of MarClean ionic cleanliness test results as compared to the Benchmark results. This material can be used by immersion, spray or other forms of agitation. It has a high flash point and a low odor. Due to the high freezing point (44°F) the material should be stored above 50°F. Customers of MarClean are advised to evaluate component compatibility prior to production use. At press time this material was available in developmental quantities from Martin Mariette Magnesia Specialities Inc. Pending additional testing MarClean will be available for commercial sale.

In May, 1990 the EPA/DOD/IPC AdHoc Solvents Working Group completed testing of MarClean R and indicated that the material met the criteria established by the Group for approval as an acceptable CFC alternative for removal of rosin solderpaste and rosin flux from electronic subassemblies. Detailed reports are available from the material manufacturer and the IPC.

5.3.5. Use of Semi-Aqueous Materials in Military Electronics

A test program was jointly designed by the EPA/DoD/IPC Ad Hoc Working Group to evaluate alternative cleaning materials. This program consisted of establishing a benchmark criteria for the cleaning performance of a CFC 113/methanol stabilized azeotrope. Alternatives are being tested and compared to the benchmark criteria in the presence of a Test Monitoring and Validation Team (TMVT).

Once the TMVT has reviewed the completed report the test data will be submitted to the Department of Defense Soldering Technology Working Group (DODSTWG). This group is responsible for reviewing the reports and implementing acceptable alternative materials into the military specifications.

REFERENCES

1. Bulat, T. J. "The Role of Cavitation in Sonic Energy Cleaning," in *Cleaning and Materials Processing for Electronics and Space Apparatus,* Special Technical Publication No. 342, American Society for Testing Materials, 1962.
2. Bulat, T. J. op. cit.
3. "Ultrasonic Cleaning for Military PWAs," Electronics Manufacturing Productivity Facility Technical Brief 008, Aug. 1989.
4. Bulat, T. J. Op. cit., p. 122.
5. EMPF. Op. cit.
6. Bulat, T. J. Op. cit., p. 125.
7. Bulat, T. J. Op. cit., p. 125.
8. McQueen, D. H. "Frequency Dependence of Ultrasonic Cleaning." *Ultrasonics,* **24,** 274 (Sept. 1986).
9. Dussault, J. "Surface Mount Cleaning Utilizing MicroCoustic® Technology, The Branson Ultrasonics Corp." Presented at NEPCON West 1989.
10. EMPF. Op. cit.
11. Goepfert, S. "Cleaning Printed Wiring Assemblies with Isopropyl Alcohol: An Attractive Alternative to Conventional Cleaning Solvents," Honeywell Inc., Presented at the Seventh Annual Soldering Technology and Product Assurance Seminar held 24–26 Feb. 1983.
12. Rich, R. "A New Technology for Cleaning Printed Circuit Boards and Hybrid Circuits," Surface Mount '89 Proceedings, San Jose, California, August 1989, pp. 551–562.

13. Archer, W. L., and T. D. Cabelka. *EXPO SMT '87 Technical Proceedings*, Las Vegas, Nevada, October 1987,pp. 127–131.
14. Cabelka, T. D. and W.L. Archer. "Cleaning: What Really Counts." Dow Chemical Co.
15. Keeler, R. "Post-Solder Cleaning Meets its Match in SMT Geometry," *Electronic Packaging and Production*, 86–89 (January 1987).
16. "Solvents Flush Solder Flux from Tight Spaces," *Electronic Packaging and Production*, 59 (August 1984).
17. Comerford, M. "Cleaning Printed Wiring Assemblies: The Effects of Surface Mounted Components," *Electri-Onics*, 13–19 (November 1984).
18. Ostrovsky, M. V. "Theoretical Foundations of Post-Solder Cleaning Process for Printed Wiring Assemblies." *Colloids and Surfaces*, **33**, 259–277 (1988).
19. Washburn, E. W. *Phys. Rev. Ser. 2*, **17**, 273 (1921).
20. Adamson, A. W. *The Physical Chemistry of Surfaces*, 2nd Edition, Chapter 10, New York: John Wiley and Sons, 1967.
21. Eichorn, J. *Petroleum Refiner*, (November 1955).
22. Burgoyne, J. H. *Inst. Chem. Eng. Symp. Ser. 15. Proc. 2nd Symp Chem Process Hazards Spec. Ref. Plant Des.* 1–5 (1963).
23. Bodurtha, F. T. *Industrial Explosion Prevention and Protection.* New York: McGraw-Hill, 1980, p. 9.
24. Zabetakis, M. G. *U.S. Bur. Mines Bull. 627* (USNTIS AD-701 576), 1965.
25. Burgoyne, J. H. *Chem. Eng. Prog.* **53** (3), 121M—124M (1957).
26. Brox, M. "The Status of Substitute Solvents and Processes to Replace CFC-113 in the Cleaning of Printed Circuit Boards at Northern Telecom." Presented at the 3rd International SAMPE Electronics Materials & Processes Conference, Los Angeles, California, June 20–22, 1989.
27. Kesevan, S., and F. Riza. "Substitutes and Alternatives to CFC-113 Based Cleaning—An Overview." Presented at the 3rd International SAMPE Electronics Materials & Processes Conference, Los Angeles, California, June 20–22, 1989.
28. Hayes, M. E. "Chlorinated and CFC Solvent Replacement in the Electronics Industry: The Terpene Hydrocarbon Alternative." *NEPCON East '88 Proceedings*, Boston, Massachusetts, June 1988, pp. 371–395.
29. Wegner, G. M. and G. C. Munie. "Defluxing Using Terpene Hydrocarbon Solvents." IPC Technical Paper No. 678.
30. Attalla, G. "Update on In-Line Cleaning with Terpene Solvents." *Surface Mount '89 Proceedings*, San Jose, California, August 1989, pp. 533–537.
31. Hayes, M. E. "Waste Minimization and High Performance Cleaning Using Semi-Aqueous Processes." *NEPCON East '89 Proceedings*, Boston, Massachusetts, June 1989, pp. 301–310.
32. Hayes, M. E. "High Performance Cleaning with Non-Halogenated Solvents." *NEPCON West '89 Proceedings*, Anaheim, California, March 1989, pp. 536–547.
33. Hayes, M. E. "High Performance Cleaning with Terpene/Surfactant Mixtures." *Surface Mount '88 Proceedings*, Marlborough, Massachusetts, August 1988, pp. 481–497.
34. Hayes, M. E. "Cleaning SMT Assemblies Without Halogenated Solvents," *Surface Mount Technology*, 35–45 (December 1988).
35. Hayes, M. E. "Naturally Derived Biodegradable Cleaning Agents: Terpene-Based Substitutes for Halogenated Solvents." *HazMat West '88 Conference, (Long Beach, California, November 1988) Proceedings*, Tower Conference Management Co., pp. 35–45.
36. Dickinson, D. A., L. A. Guth, and G. W. Wegner. "Advances in Cleaning of Surface Mounted Assemblies." *NEPCON West '89 Proceedings*, Anaheim, California, March 1989.

37. Hamblett, G. W., and G. A. Larson. "Terpene/Aqueous Cleaning," *NEPCON West '90 Proceedings*, Anaheim, California, February 1990.
38. Tashjian, G., G. Wegner, and D. Dickinson. "Terpene Cleaning Implementation for SMT." *NEPCON West '90 Proceedings*, Anaheim, California, February 1990.
39. Elliot, D. "Is Aqueous the Answer? *NEPCON West '90 Proceedings*, Anaheim, California, February, 1990.

6
Defluxing for High Reliability Applications and General Environmental Issues

Joe R. Felty
Texas Instruments, Inc.

6.1 INTRODUCTION

Cleaning electronic hardware to remove contaminants resulting from handling and assembly operations has long been a standard practice. The level of cleaning required often is governed by aesthetic (visual), electrical test, and performance requirements. Several of the criteria/requirements discussed herein relate to the manufacture of military electronic hardware, however, many are pertinent to the production of any high quality/high reliability equipment regardless of the customer or end-item use. The materials and processes used are governed by the type of product to be cleaned, the contaminants being removed, and the compatibility of the cleaning system with the product and the environment.

6.2 BACKGROUND

In the field of high performance, long-life electronics the need for extended levels of cleanliness is essential. Such equipment is often required to function for extender periods in adverse environments. Mission critical (avionics controls equipment, missile electronic systems, etc.) or life support (pacemakers, etc.) equipment requires a heightened level of reliability with extended periods of failure free performance. Customers and manufacturers of military, space and medical electronics equipment have traditionally required continuous increases in equipment performance and decreases in packaging size resulting in increased component densities on printed wiring assemblies. These increasing packaging densities have further increased the cleaning challenges of contaminant removal and the verification of the level of cleanliness attained.

6.3   CLEANLINESS REQUIREMENTS

Electronic equipment requiring high reliability, low maintainability presents a challenge to both design and assembly functions. The primary contaminants to be removed during electronic assembly are handling soils (body oils, finger print oils and salts, etc.) and residues from fluxes (currently only rosin based fluxes per MIL-F-14256 are allowed by military electronics assembly standards and specifications) used in the soldering operations. The electronic design must be inherently sound and is often dependent on the electrical layout of the circuit board and quality of electronic components specified for the assembly. However, of equal importance is the inclusion of proper mounting criteria into the design to allow for adequate spacing betwen components and the board substrate to aid in adequate removal of contaminants introduced by the electronics assembly process(es). (Refer to Chapter 1 for more details on design and component layout considerations and how they impact cleaning.)

6.3.1.   Necessity for Cleanliness

Without adequate removal, the soils introduced during assembly processing can provide current leakage paths to develop between adjacent board circuitry or component leads resulting in decreased system performance or in some cases, direct electrical shorts which can produce catastrophic failure of the system. The conductive nature of the contaminants themselves or more often their hygroscopic nature will result in moisture accumulation which can provide the path for electronically conductive mobile chemical species to transfer electrical currents. Therefore, the ability to remove contaminants from both a design and manufacturing standpoint is critical and synergistic.

The presence of contaminants with the potential for causing current leakage/shorting problems can also generate reduced system performance or failure as an aftermath of corrosion reactions. Materials such as free chlorides, bromides, etc. can participate in chemical reactions with the basis metal of board circuitry or component leads. These corrosive reactions if allowed to proceed for a sufficient length of time can produce a substantial decrease in overall circuit cross-sectional area resulting in a conductor trace that cannot conduct the required current load. This decrease can create excessive over heating of the circuitry or lead during operation and a subsequent complete deterioration or break in the conductor, forming an open circuit.

The removal of assembly contaminants is also beneficial in subsequent visual inspection of overall assembly quality. The presence of flux residues, for example, can obstruct or completely mask the viewing of solder joints. These residues can also interfere with postassembly electrical testing by contaminating test fixture connections resulting in erroneous readings due to the insulating or conductive nature of the particular residues in question.

In many high reliability, high performance electronics, the "conformal" coating of printed circuit assemblies with a polymeric material is required as a final step in the manufacturing operation. The coating serves several purposes, one of which is encapsulation of contaminants (flux, solder splatter, wire clippings, etc.) that might inadvertently remain on the board. Embedment of contaminants in the polymer matrix can prevent the residues from migrating during subsequent operations, thereby minimizing the opportunity for electrical shorts, etc. The coating can also prevent the intrusion of liquid water and some solvents down to board circuitry or component leads, thereby preventing shorting between adjacent conductors.

However, the presence of many organic and inorganic residues beneath the coating can result in inadequate adhesion of the coating to the board, component, lead or conductor surfaces. The loss of adhesion can produce a visual condition known as mealing where white discolorations occur at the coating-board substrate interface. These areas of adhesion loss provide sites for moisture accumulation and corrosion and can result in degradation or loss of electrical performance if they encompass or bridge circuitry. Therefore, the need for contaminant removal is varied and can be critical to system electrical performance and longevity. (Refer to Chapters 3 and 4 for more details of the effects of contaminants.)

6.3.2.   Military Requirements

The military has long recognized the need for effective cleaning measures in electronics assembly. The original attempt to provide a quantifiable determination of cleanliness levels was documented in MIL-P-28809. The cleanliness requirements for this specification were based on work published in 1972 by the Naval Avionics Center.[1] Using a solvent extract resistivity test, the upper limit of contamination as determined by the presence of ionic species soluble in a 75% isopropyl alcohol (IPA)/25% water solution was set at 10.06 micrograms of sodium chloride (NaCl) equivalent/square inch of surface area tested. Although several pieces of automated equipment have been developed over the ensuing years to augment the original manual test technique, the limit of 10.06 micrograms/square inch has remained as the maximum ionic contaminants allowed. An obvious inadequacy of this method centers around the fact that only alcohol/water soluble ionic species are detected.

Organic materials that are not soluble in water/alcohol solution or those that are soluble but are not ionic in nature will not be affected. Visual inspection is used primarily to detect the presence of organic residues on hardware. Test methods for qualitatively and quantitatively determining nonionic cleanliness levels are available, but in general are not as suitable for use in the manufacturing/assembly area as is the solvent ionic extract test. A test method involving applying acetonitrile to an assembly and collecting the runoff on a glass slide, allowing the solvent to evaporate and subsequent visual inspection for evidence of organic residues is available.[2] A further refinement involves collecting the run off followed by quantitative identification of nonionic residues using infrared spectrophotometric analysis.[3] More recently a quantitative test using IPA and analytical examination of the IPA extract solution by UV-VIS spectrophotometry appears to be quite sensitive and repeatable.[4]

Although the ionic solvent extract resistivity test does not detect the presence of nonionizable contaminants, it has had wide spread use throughout the military contractor industry since 1975. The preponderance of evidence indicates that where the test is applied, on a continuing basis, assemblies meeting this requirement before conformal coating or before installation in the next higher assembly have not experienced field failures that were attributable to board contaminant problems. Therefore, indirectly, assemblies that can pass the ionic cleanliness test would appear in most cases to be sufficiently free of organic residues such that quality and reliability are not adversely affected.

6.4 CLEANING MATERIALS

Numerous cleaning materials are available for cleaning electronic assemblies. All sectors of the electronics industry have had a long and successful history in removing assembly residues from printed wiring boards and assemblies using petroleum based solvents[5] and aqueous cleaners[6] individually or in tandem.

6.4.1. Solvents

Initially in the early 1960s hydrocarbons such as naphtha and other petroleum distillates or alcohols such as isopropyl alcohol, etc. were used to remove assembly flux and handling oil residues. However, the inherent safety issues (flammability) associated with such solvents drove the industry to pursue other cleaning options. In the 1970s, the use of chlorinated solvents such as 1,1,1-trichloroethane (methyl chloroform) came into popular use. Methyl chloroform provided a medium that was not flammable in normal use and was more aggressive in solubilizing organic residues though less effective in removing ionic residues. The material, through the use of distillation equipment, provided for user recovery and reuse of the cleaning solvent. However, since the cleaner was more aggressive, there were at times compatibility problems between the solvent and the material constituents of the assemblies being cleaned. In addition worker exposure to this hydrochlorocarbon solvent and its vapors had to be controlled due to the moderate threshold limit values (TLVs) established for maximum worker exposure. The TLV for methy chloroform is 350 ppm in air.

In the late 1970s and early 1980s came the emergence and wide spread use of chlorofluorocarbon (CFC) cleaning solvents. These solvents (trade names such as Freon, Genesolv, Arklone, etc.) and their alcohol azeotropic mixtures provided not only organic solvency provided by the chlorofluorocarbon, but also ionic solvating properties due to the presence of alcohols (i.e. isopropyl, n-propanol, methanol, etc.). In addition, worker exposure concerns over contact with the solvent or vapors were further lessened because of the relatively low toxicity of the CFCs (TLV of 1000 ppm in air for the solvent 1,1,2-trichloro-1,2,2-trifluoroethane [CFC-113]) and the nonflammable nature of the solvents including their alcohol azeotrope mixtures. Another advantage of these solvents is their compatibility with most materials used in electronic assemblies. This compatibility is primarily attributable to the inherent stability of the chlorofluorocarbon molecule. As will be mentioned later this stability has led to other environmental concerns, which may eventually lead to worldwide cessation of production and availability of the materials in the near future.

6.4.2. Aqueous

Aqueous cleaning material/process alternatives became readily available for the postsolder cleaning of electronic assemblies in the mid- to late 1970s. Solutions of water and saponifiers (primarily ethanolamines) at elevated temperatures (normally 140–160°F were used to react with organic residues (for the most part rosin flux residues) to form water soluble soap byproducts. The byproducts could then be removed using subsequent rinsing with copious quantities of water. These materials provide high worker safety in the concentrations used in electronics assembly cleaning and are considered biodegradable. The concentration of saponifiers must be controlled to prevent damage to assembly part making during clenaing. Saponifiers, most of which are basic, must be adequately removed during the cleaning operation or they may in turn become corrosive media on the assembly.

Aqueous cleaning without saponifiers is also used for cleaning of assemblies processed with nonrosin water soluble fluxes. In this condition, neutralizers or rinse aids are added to the aqueous solution to remove the water soluble constituents of the fluxes. The use of aqueous cleaning in commercial electronics assembly where water soluble fluxes have been employed has proved successful for several years. In either case, the removal of rosin or nonrosin water soluble fluxes, requires large quantities of water. (Refer to Chapter 4 for additional information on aqueous cleaning.)

6.4.3. New Alternatives

Recent concerns over the environment and the negative impact of some chlorofluorocabon compounds with the stratospheric ozone layer of the earth's atmosphere have led to the rapid development of new alternatives for electronic assembly cleaning. One of the new alternatives, the use of naturally occurring organic compounds derived from pine (terpene) and citrus products, has gained rapid notoriety. The active ingredient, primarily limonene, has been shown to be a particularly good solvent for flux residues resulting from soldering operations.[7,8] These materials are applied in the cleaning operation similar to previously mentioned aqueous cleaning operations. For best results, the terpene solution is sprayed on printed wiring assemblies followed by rinsing with large quantities of water to emulsify the solution and remove it along with the dissolved contaminants from the board. This cleaning technology has also been labeled as, "semi-aqueous" cleaning since the technique requires an organic solution for solvating contaminants followed by aqueous rinsing.

New entries into the CFC alternatives approach to cleaning printed circuit assemblies involve partially halogenated CFCs, where one or more chlorine or fluorine molecules of a fully halogenated CFC have been replaced by a hydrogen atom. Such materials are more commonly referred to as HCFCs (Figure 6.1). As a result of the presence of at least one hydrogen atom these materials result in a molecule that is not as stable as pure CFCs. Therefore, these materials in the vapor state have shorter atmospheric lifetimes when emitted. HCFCs have much lower ozone depleting potentials (ODP) than CFC-113 (typically 0.05 ODP compared to 0.80 ODP) because of this inherent relative instability. Long-term toxicology testing to assess industrial hygiene impact has not been completed on the HCFC solvents at the time of this writing and therefore their future potential as a CFC-113 replacement and availability in large commercial quantities is still unknown. If these materials are found to be acceptable from a toxicology standpoint, they could possibly be replacements for CFC solvent cleaners. However, new or existing defluxing equipment will require design modifications (smaller openings to the atmosphere, increased vapor condensation zones, etc.) in order to adequately retain these solvents to compensate for their lower solvent boiling temperatures (higher vapor pressures). Some minor maintenance to incorporate solvent resistance pump seals may also be required.

6.4.4. Acceptance of Alternatives

The commercial electronics industry has a history of rapid incorporation of new technologies as they are developed, i.e., water soluble fluxes, low residue fluxes, no clean fluxes, surface mount technology, etc. There is no reason to think that this will not be the case for the new CFC alternative cleaning materials. Once

CFC-113:

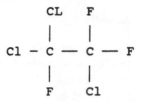

(1,1,2-trichloro-1,2,2-trifluoroethane)

HCFC-141b:

(1,1-dichloro-1-fluoroethane)

HCFC-225ca:

(1,1-dichloro-2,2,3,3,3-pentafluoropropane)

Fig. 6.1. Structure of CFC-113 versus HCFC solvents.

equipment is readily available to safely handle spray applications of the terpene cleaners in the electronics assembly area and the toxicity data (if acceptable) for HCFCs is completed, their utilization should rapidly increase.

The military sector of the electronics industry has historically taken a more conservative approach to the incorporation of new cleaning technologies. One of the more familiar specifications in the area of military high reliability electronics assembly is WS-6536, "Procedures and Requirements for Preparation and Soldering of Electrical Connections." This widely used document allows only specific solvents, primarily isopropyl alcohol, methyl chloroform, and 1,1,2-trichloro-1,2,2-trifluoroethane (trichlorotrifluoroethane or CFC-113) along with

Table 6.1. Cleaning Solvents Allowed by Military Specifications and Standards.

WS-6536E

3.3.3 Solvents and cleaners. The solvent or cleaner used for removal of grease, oil, dirt, flux, and other debris shall conform to Table II.

TABLE II. Solvents and Cleaners.

SOLVENT	SPECIFICATION
Ethyl Alcohol	O-E-00760, Types III, IV or V
Isopropyl Alcohol	TT-I-735
Methyl Alcohol	O-M-232, Grade A
N-Propyl alcohol	ASTM D 3622
Butyl Alcohol, Secondary	TT-B-848
1, 1, 1-Trichloroethane	MIL-T-81533
Trichlorotrifluoroethane	MIL-C-81302
Trichlorotrifluoroethane	MIL-C-85447, Type II
Solvent Petroleum	Drawing 200AS311
Distillate (Stoddard)	
CLEANERS	SPECIFICATION/NOTE
Reagent water (Type II)	ASTM D 1193
Detergent cleaners	As approved by the Government procuring activity

Source: WS-6536E, SCN-4 (Issued 14 February 1989)

---

its alcohol azeotropes, to be used in electronics assembly cleaning operations (Table 6.1). The use of other cleaning materials on hardware assembly to the requirements of this specification requires specific approval by the procuring activity on a contract by contract basis. Aqueous cleaning incorporating the use of saponifiers, readily available since the late 1970s, still requires approval prior to use per this document.

Recent environmental concerns over the effects of CFC (and possibly methyl chloroform) emissions on the stratospheric ozone layer and the Department of Defense (DOD) commitment to reduce dependency on such materials may have been in part responsible for a change in philosophy when defining assembly cleaning requirements. The latest standard addressing the requirements for military electronics assembly is MIL-STD-2000, "Standard Requirements for Soldered Electrical and Electronic Assemblies." This document represents a major cooperative effort by both the military and industry to generate a set of electronics assembly criteria acceptable to all of DOD that are directed toward end-item

Table 6.1. (Continued)

MIL-STD-2000

4.10.4 *Solvents and cleaners*. The solvents or aqueous cleaners used for removal of grease, oil, dirt, flux, and other debris, shall be selected for their ability to remove both ionic and non-ionic contamination. The solvents or cleaners used shall not degrade the materials or parts being cleaned. A list of approved solvents and cleaners is provided in tables I and II. If other solvents and cleaners are used, analysis and documentation demonstrating compliance with the above requirements shall be available for review and disapproval. Mixtures of the approved solvents may be used.

TABLE I. Solvents.

SOLVENT	SPECIFICATION
Ethyl Alcohol	O-E-760, Types III, IV or V
Isopropyl Alcohol	TT-I-735
Methyl Alcohol	O-M-232, Grade A
Butyl Alcohol, Secondary	TT-B-848
1, 1, 1-Trichloroethane	MIL-T-81533
Trichlorotrifluoroethane	MIL-C-81302
Trichlorotrifluoroethane	MIL-C-85447, Type II
Solvent Petroleum Distillate (Stoddard)	Use Appendix B

TABLE II. Cleaners.

CLEANERS	SPECIFICATION/NOTE
Water	1 megohm-cm, minimum resistivity
Detergent cleaners and saponifiers	Subject to review and disapproval

Source: MIL-STD-2000 (Issued 16 January 1989)

hardware requirements. In that respect, the requirements for cleaning of hardware have taken an innovative approach. The same list of preferred solvents is present in MIL-STD-2000 as is in WS-6536 (Table 6.1). However, MIL-STD-2000 allows the use of other cleaning materials without prior customer approval provided the contractor has available for review and possible disapproval objective data indicating that the cleaning material/system employed does not detrimentally impact the product being cleaned and that the cleanliness requirements specified by the standard are maintained. In addition, the use of aqueous cleaning is now allowed without the need to obtain prior approval. The requirement as now stated in MIL-STD-2000 will allow the rapid implementation of new alternative cleaning materials as they become available, provided assembly cleanliness requirements are met, without the need to revise the document.

6.5.   CLEANING EQUIPMENT

Two techniques, manual and mechanized, are prevalent in electronics cleaning. With ever increasing component densities and decreasing spacing between components and component circuitry, manual cleaning has been relegated to an interim operation during manual assembly and soldering of hardware where the formation of electrical interconnections is not suitable to mass soldering techniques. When high levels of hardware cleanliness are paramount, mechanized cleaning operations involving sophisticated equipment to apply the cleaning media provide the standard for cleaning today's high performance commercial, medical, and military electronics communication and life support equipment.

6.5.1.   Batch Cleaners

The ability of solvent cleaners to adquately remove both organic and especially ionic contaminants is enhanced if the solvent also possesses a constituent that has an affinity for ionic soils (such as polar alcohol, etc). A basic approach to solvent cleaning involves vapor batch defluxing equipment similar to that in Figure 6.2. Such systems if adequately maintained can provide distilled solvent and solvent vapor cleaning of a sufficient quality level to meet military and medical high reliability cleaning requirements. With the addition of sliding lids or covers integrated with a hoist system, the emission losses of solvent to the atmosphere can be greatly reduced.

Batch cleaners are also available for aqueous cleaning applications. Such systems which include a programmable cleaning cycle in conjunction with an interactive resistivity monitor can provide a system for maintaining and controlling a high level of product cleanliness. The continuous monitoring of rinse water resistivity after completion of the heated saponifier water wash cycle can assure adequate wash/rinse cycle times sufficient to remove water soluble ionic contaminants to some preselected cleanliness level (Figure 6.3). Hardware cleaned in this manner should easily pass ensuing solvent extract resistivity testing referenced previously in Section 6.3.2.

6.5.2.   Inline Cleaners

The most popular solvent and aqueous cleaning equipment for medium to high volume printed wiring board assembly lines are inline solvent and/or aqueous cleaners placed in the assembly flow immediately afer soldering. The latest design concepts incorporated into inline solvent cleaners involving liquid seal traps at

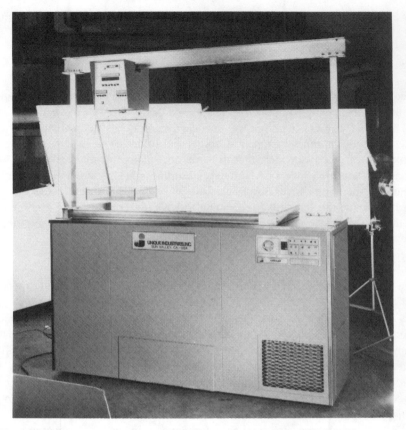

Fig. 6.2. Typical batch solvent vapor defluxer (Vapo-Kleen series vapor degreaser with auto-arm hoist, courtesy of Unique Industries).

exit and entrance openings below the hot vapor level, and increased free-board design have greatly reduced solvent losses to the environment (Figure 6.4). In addition, the solvent traps allow higher volume throughput and higher spray pressures in the cleaning sections. The high pressures provide increased solvent displacement potential in tight geometries and the area(s) beneath and between the component body and mounting substrate. Were it not for the solvent traps, the turbulence in the vapor blanket produced by the high pressure spray would severely limit the use of such a cleaning technique because of high solvent vapor loses.

Aqueous inline cleaner designs over the years have evolved to units which expose parts to high volume moderate pressure spray or jet applications of aqueous or aqueous/saponifier cleaning solution (Figure 6.5). Normally, the key

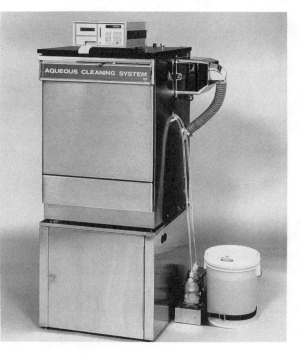

Fig. 6.3. Batch aqueous cleaner. (Model 6300 uP aqueous cleaning system, courtesy of Electronic Controls Design, Inc.)

to the success of such systems in meeting military cleanliness requirements involves the judicious use of high velocity air knives and multiple wash and rinse sections in the equipment design. Since aqueous removal of rosin fluxes requires the use of a saponifier (caustic in nature), after application to the board the materials must be thoroughly removed not only to remove the saponified or solvated soils but also to remove any residual cleaning agent, which if allowed to remain could pose a possible corrosion potential. Multiple wash sections separated by air knives assure that the majority of cleaning solution after each application is removed from tight geometries and from beneath components. This action not only removes contaminants but also provides dynamic forces to purge cleaning solution from the board and beneath components. In addition, fresh cleaning solution from subsequent cleaning sections is allowed to penetrate into tight geometries to complete the saponification/solvation of unwanted residues. A similar combination of water rinse/air knife sections immediately after the wash sections can assure that contaminants including residual cleaning solution are adequately removed to meet preestablished cleanliness requirements.

Since organic solvent and aqueous cleaning materials are not universal solvents

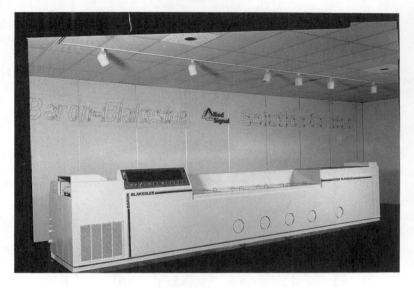

Fig. 6.4. Typical conveyorized inline solvent liquid/vapor defluxer. (Model CBL-LE defluxer, courtesy of Allied Signal, Inc./Baron-Blakeslee.)

Fig. 6.5. Typical conveyorized inline aqueous cleaner. (Model 500 Hydrocleaner, courtesy of Electrovert USA Corporation.)

for both organic and inorganic contaminants, a combination of solvent cleaning followed by aqueous cleaning has been required in many circumstances to consistently meet MIL-P-28809 cleanliness criteria on military electronics printed wiring assemblies (PWA). This combination of cleaning techniques has been exceptionally successful, especially if a wide variety of product designs and configurations are processed through the same soldering/cleaning equipment and processes. Numerous articles have been published on this subject, but one of the earliest was generated by the Navy.[9] This is not to say specific designs cannot be cleaned to the MIL-P-28809 or other more stringent cleanliness levels by either solvent only or aqueous saponifier only cleaning; it will merely be more difficult. Where assembly lines can be dedicated to one or two product designs, the use of a single cleaning media can be successful provided the machine and process capabilities are well understood and adequate process controls are instituted and monitored. In addition, if a hydrocarbon or halogenated hydrocarbon solvent cleaning system is utilized, a blend of that solvent with a more polar constituent, such as methanol or ethanol, is essential to assure that ionic residues will be adequately removed along with the other organic soils present.

6.6.   CLEANLINESS VERIFICATION

As mentioned earlier, the soils or contaminants present on PWAs after soldering and assembly can be both organic or inorganic in nature. And, as was the case with the removal of contaminants, to determine the presence and level of contaminants on the PWA different test techniques and test equipment or test media must be used. Also as was formerly stated a generally accepted industry standard for electronics assembly cleanliness level was established by MIL-P-28809. The specification allows for both manual and automated test procedures to determine ionic cleanliness levels.

6.6.1.   Ionic Test Methods

The original manual solvent extract resistivity method for determining ionic contaminant levels has subsequently been replaced by commercially available equipment capable of conducting the test procedure and calculating resultant cleanliness levels (Figure 6.6). The equipment utilizes a test solution of 75% isopropyl alcohol/25% water. The alcohol has the capability of solubilizing most organic residues present on the assembly thus freeing any ionic contaminants that might be covered/complexed in the organic matrix (i.e., rosin flux residues could possibly encapsulate ionic soils and prevent their solvation or dissolution in the aqueous solutions.) The mobile ionic species are preferentially soluble in the more polar water solution and once mobile are capable of being quantitatively determined by measuring the change in resistivity of the test solution before and

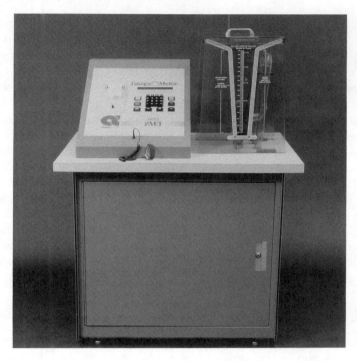

Fig. 6.6.   Instrument used to determine ionic contaminant levels on printed wiring assemblies. (Omega Meter 600 SMD, courtesy of Alpha Metals, Inc.)

after exposure to the PWA to be evaluated. The maximum allowable contaminant level of 10.06 micrograms of NaCl equivalent/square inch of surface area tested was baselined using the manual method. Subsequently, equivalency factors for commercially available test equipment have been developed which relate back to the manual method.[10] The equivalency factors (Table 6.2) compensated for the variances between several automated test equipment designs based on their detection of constant levels of contamination on printed wiring assembly test coupons. The "Acceptance Limits" for automated equipment were consistently higher than the manual method.

Another method of determining cleanliness levels involves surface insulation resistance (SIR) testing. SIR meaures the resistance of the surface of a dielectric material between two parallel conductors. The test indirectly measures the presence of ionic or mobile contaminants betwen the parallel conductors by monitoring current leakage induced by the contaminants. Several papers have been published on the subject[11,12,13] and attempts have been made to correlate SIR test results with the previously mentioned solvent extract resistivity test.[14] Although a sensitive test, the presence of nonmobile, nonionic residues are not

Table 6.2. Equivalency Factors for Automated Test Equipment[7].*

METHOD/EQUIPMENT	$\overline{\overline{X}}$, µg NaCl/in.^{2}	EQUIVALENCY FACTOR	INSTRUMENT (ACCEPTANCE LIMIT)	
			µg NaCl/cm^2	µg NaCl/in.^{2}
MIL-P-28809[a] (Beckman)	7.47 $\overline{\overline{X}} = 7.545$	$\dfrac{7.545}{7.545} = 1$	1.56	10.06
MIL-P-28809[a] (Markson)	7.62	$\dfrac{7.545}{7.545} = 1$	1.56	10.06
Omega Meter	10.51	$\dfrac{10.51}{7.545} = 1.39$	2.2	14
Ionograph	15.20	$\dfrac{15.20}{7.545} = 2.01$	3.1	20

*Table amended.
[a]Naval Avionics Center (NAC) modified MIL-P-28809 test utilizing commercially available test equipment: Beckman Conductivity Bridge (Model RC-16C) and Markson ElectroMark Conductivity Analyzer (Model J-4405).

detectable. The SIR test requires an extensive amount of electrical and environmental test equipment as well as a high degree of technical skill for precise results. Equipment is now available (Figure 6.7) which provides a method of incorporating the various electrical and environmental control requirements in one package thereby simplifying testing logistics.

6.6.2.   Organic Residue Test Methods

The other major category of contaminants most often present on PWAs are organic residues from handling (primarily body oils, particulate matter, etc.) and electronics assembly processing (flux and flux residues, machine oils, etc.). Although if present at the time of ionic testing and capable of being dissolved in the alcohol/water solution, such materials would not be detected by the solvent extract resistivity test since this test is sensitive only to ionic residues. The detection of organic residues is possible; however, commercially available equipment currently does not exist that will provide an operator friendly method adaptable to the electronics assembly area for quantitatively determining the presence of organic residues.

The quantitative detection of the primary residual organic contaminant (rosin) resulting from electronics soldering has been reported[15,16] and is also a laboratory test. This test method involves extracting residual contaminant with a specific solvent, then utilizing UV-visible spectrophotometric methods to quantify the

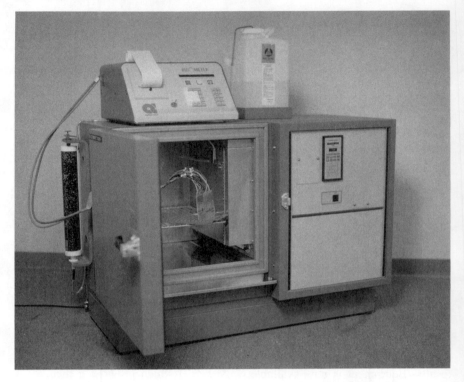

Fig. 6.7. Surface insulation resistance test equipment—integrates both microprocessor controlled environmental chamber and surface resistivity measurement equipment. (Sir-O-Meter, courtesy of Alpha Metals, Inc.)

amount of rosin present. Again the success of this test technique is heavily dependent upon the extraction technique and technical skill of the laboratory technician.

The further commercialization of equipment for conducting operator friendly organic and SIR testing would be desirable especially in high density complex geometry (surface mount technology) applications where hardware reliability is a premier requirement. When high reliability testing is necessary then the more time consuming, increased operator skill level, measurement of residual organic residue analysis or surface insulation resistance testing may be cost effective.

In depth discussion of all the test methods mentioned can be found in both Chapter 3 and Chapter 4 of this book.

6.7. ENVIRONMENTAL ISSUES

The materials used in the assembly of electronics and their impact upon worker safety and the environment are issues of major signifciance. Whether emitted to

Table 6.3. Atmospheric Lifetimes of Montreal Protocol CFCs.

CHEMICAL	FORMULA	LIFETIMES (yrs)
CFC-11	CCl_3F	60
CFC-12	CCl_2F_2	120
CFC-113	CCl_2FCClF_2	90
CFC-114	$CClF_2CClF_2$	200
CFC-115	$CClF_2CF_3$	400

the atmosphere, dispersed through effluent streams or collected and disposed of by incineration, land-fill or deep well isolation, the final disposition of waste byproducts of cleaning oeplrations are being monitored and controlled with ever increasing scrutiny.

The limits on and type of worker exposure to chemicals used in the electronics industry as well as other industries are well documented and controlled by Occupational Safety and Health Administration regulations. To assure continued worker safety, new materials proposed for such uses must successfully pass a series of strenuous, often lengthy evaluations prior to being made commercially available.

6.7.1. Atmospheric Impact

Similar U. S. Environmetnal Protection Agency (EPA) regulations are also in place governing emission and disposal of these chemicals. For example the emission of volatile organic compounds (VOCs) that are photochemically reactive, producing such irritants as ozone, smog, etc. in the lower troposphere are closely regulated. These regulations fall into the realm of federal, state and local jurisdiction and are intended to provide industry with quantifiable requirements to assure the protection of the environment. Until recently, the emissions of chlorofluorocarbons (CFCs) were thought, because of their chemical stability, to be nonthreatening to the environment. However, mounting search for evidence over the last three to four years has virtually proven that CFCs are the primary factors behind the destruction of the earth's stratospheric ozone (O_3) layer. The very property (chemical stability) that makes the fully halogenated CFCs so attractive and results in low human toxicity and relative chemical inertness/compatibility with most materials is the primary reason for their detrimental impact on the ozone. The long chemical life (Table 6.3) of these molecules allows for the time required to transport them to the upper layers of the atmosphere. Once there, exposure to the sun's ultraviolet (UV) radiation degrades the compounds, liberating active chlorine species which react catalytically with ozone eventually breaking it down to oxygen. In addition, neither chlorine nor oxygen have the UV protecting properties of ozone, and EPA projections indicate that for every 1% prolonged decrease in stratospheric ozone a corresponding 3%

Table 6.4. Montreal Protocol (September 1987) Terms of Agreement.

Chlorofluorocarbons (CFC): -11, -12, -113, -114, -115	
Freeze at 1986 production levels	July 1, 1989
20% reduction from 1986 levels	July 1, 1993
50% reduction from 1986 levels	July 1, 1998
Halons: -1211, -1301, -2402	
Freeze at 1986 production levels	July 1, 1992

increase in skin carcinomas in the northern hemisphere may result. Other adverse results of increased UV radiation reaching the troposphere involve possible suppression of the human immune system and perturbation of life cycles of microorganisms which form the basis of the food chain.

6.7.2. Montreal Protocol

As a result of these health and environmental issues, the United States Environmental Program met in 1987 to discuss the global impact of ozone depletion, its probable causes, and corrective actions. The result was the Montreal Protocol[17] of September 1987, which initially ratified by 24 countries, proposed a timed global reduction of specific CFCs and Halons (bromofluorocarbons). At its maximum impact (Table 6.4) the protocol would require a 50% reduction in global CFC production by 1998 based on 1986 production levels. The protocol was entered into force on 1 January 1989 and the first prescribed reduction, that of decreasing world CFC production back to 1986 levels, went into effect 1 July 1989. The CFCs affected (Table 6.5) are fully halogenated carbon based molecules containing at least one chlorine atom. The Halons (primarily used as fire extinguishing materials) contain both chlorine and bromine and as Table 6.6 indicates are molecule per molecule higher ozone depleters than the CFCs.

As of this writing, data continue to accumulate indicating the destruction of the ozone layer is progressing faster than originally predicted. Further reduction of ozone depleting chemicals may be required above and beyond the current Montreal Protocol limits if the stratospheric levels of ozone depleting chemicals, primarily chlorine monoxide (ClO), are to be significantly reduced back to 1975 levels before the year 2100. The findings of the Ozone Trends Panel Report of 1988[18] disclosed that CFC stratospheric decomposition products may have been responsible for a 2% decrease in the global level of ozone since 1978. There are intimations that other less ozone depleting chemicals such as 1,1,1-trichloroethane (methyl chloroform) and carbon tetrachloride will be added when the protocol members are convened again in June 1990 in the United Kingdom. The member nations will consider whether or not the stringency and chemical coverage of the protocol restrictions should be increased. Although methyl chlo-

Table 6.5. CFCs and Halons under Montreal Protocol Restrictions.

CHEMICAL	FORMULA	USES
CFC-11	CCl_3F	Foam blowing, refrigeration, (chillers, aerosol propellant)
CFC-12	CCl_2F_2	Air conditioning (auto, appliances) refrigeration (chillers), foam blowing, sterilization
CFC-113	CCl_2FCClF_2	Cleaning solvent
CFC-114	$CClF_2CClF_2$	Refrigeration (chillers)
CFC-115	$CClF_2CF_3$	Refrigeration
Halon 1211	$CBrClF_2$	Fire extinguisher
Halon 1301	$CBrF_3$	Fire extinguisher
Halon 2402	$CBrF_2CBrF_2$	Fire extinguisher

roform has a lower ozone depleting potential and a shorter atmospheric lifetime than CFC-113 (Table 6.7), the production/consumption of methyl chloroform is 8–10 times that of CFC-113. Therefore, it seems that even if CFCs were phased out completely, the production of methyl chloroform will as a minimum need to be frozen at some level in order to stabilize chlorine active species at current atmospheric levels.

6.7.3. CFC Solvent Availability

The availability of CFCs in the near future will definitely be limited and depending on the results of the protocol update in June 1990 may possibly be put on a schedule requiring complete elimination by the year 2000. This action could put heavy pressure to use methyl chloroform where applicable in place of CFC

Table 6.6. Ozone Depleting Potentials (ODPs) for CFCs and Halons under Montreal Protocol Restrictions.

CHEMICAL	FORMULA	ODP
CFC-11	CCl_3F	1.0
CFC-12	CCl_2F_2	1.0
CFC-113	CCl_2FCClF_2	0.8
CFC-114	$CClF_2CClF_2$	1.0
CFC-115	$CClF_2CF_3$	0.6
Halon 1211	$CBrClF_2$	3.0
Halon 1301	$CBrF_3$	10.0
Halon 2402	$CBrF_2CBrF_2$	6.0

Table 6.7. ODPs and Atmospheric Lifetimes of CFC-113 and Methylchloroform.

CHEMICAL	FORMULA	ODP	LIFETIME*
CFC-113	CCl_2FCClF_2	0.8	>90 years
Methylchloroform	CH_3CCl_3	0.15	6.3 years

*Period of time that chlorine associated with a compound remains in the atmosphere

solvents. However, due to the potential impact on the ozone, the US EPA in August 1988 issued an Advanced Notice of Rule Making (ANPRM) in an attempt to discourage the possible wholesale conversion by CFC users from CFC solvents to methyl chloroform. The ANPRM (53 FR 30617) states "EPA expressly does not view shifting away from CFC-113 to other chlorinated solvents which are currently under regulatory scrutiny by the EPA as an acceptable solution to protecting the ozone layer."

It would appear that in terms of solvent cleaning, volatile organic solvent materials containing chlorine will not be favorably looked upon by regulatory agencies in the foreseeable future. Designing assembly cleaning systems and processes for use with chlorinated solvents in the near future would appear to have significant risk factors from a cleaning solvent availability and material acquisition cost standpoint, especially if demand increases and production levels are frozen or reduced.

In addition to the production volume restraints invoked by the Montreal Protocol, other regulatory restrictions have been placed on CFC producers. When CFC production cutback to 1986 levels was enforced in July 1989, each major CFC producer was forced to reduce production to an allocation equal to their 1986 levels. To encourage strict adherence by the protocol requirement, a fine of $5,000 will be leveled for every kilogram of CFC produced over the company's allocated limit.

Finally, from an environmental standpoint, CFCs have also been implicated as possible contributors to the global warming phenomenon. The CFCs global warming potential, i.e., their ability to prevent the radiation of thermal energy from the earth's surface through the trophosphere back outside the earth's atmosphere, is greater than that of carbon dioxide (CO_2). However, CFCs are present in much smaller concentrations in the atmosphere than CO_2 and present little threat. The burning of fossil fuels is by far the major source of generating greenhouse gases, primarily CO_2.

6.8   THE FUTURE OF CLEANING

A global commitment to protecting the atmosphere from destruction of the earth's ozone layer has obviously been initiated. Numerous unprecedented cooperative

Table 6.8. Unilateral Actions for Eliminating CFC-113 Use.

COUNTRY	GOAL	DATE
Sweden	Phase out solvent uses	1990
Germany	Phase out	1995
Canada (proposed)	Phase out for electronics cleaning	1991–94
	Phase out for metal cleaning	1990–91
	Phase out for dry cleaning	1991–92
Ontario and British Columbia (proposed)	Phase out	1991

efforts have been established on a worldwide nonpartisan basis. A voluntary coalition called the EPA/DoD/Industry Ad Hoc Solvents Working Group has been actively pursuing the evaluation of alternative materials to replace chlorofluorocarbon solvents for electronics assembly cleaning operations. The goal of the group is to coordinate and expedite the testing of alternatives using a standardized test procedure, disseminate the information on all test results throughout the electronics community, and provide the technical basis for modifying industry specifications (especially military documents) to allow use of acceptable CFC alternatives. The EPA has estimated that as much as 50% of CFC-113 consumption for cleaning of electronics assemblies is driven directly or indirectly by military specifications. The efforts of the group in developing the test procedure and subsequent results on the evaluation of CFC alternatives are being reviewed by members of industry organizations in the United States, Canada, Asia and Europe. Such international cooperation has seldom been observed. (Refer to Chapter 3, Section 3.4.2 for further information on the Ad Hoc Working Group.)

6.8.1.   Unilateral CFC Reductions

As a result of the Montreal Protocol numerous countries (Table 6.8) and major international companies (Table 6.9) have committed to time phased reduction and ultimate elimination of their dependence on and consumption of ozone depleting CFCs. Many of these reduction programs are significantly more aggressive than the existing protocol requirements and the projected increased controls that may result from the 1990 protocol update. Such actions will most assuredly impact the cleaning technologies utilized in not only the electronics industry, but also other applications in metals cleaning, precision mechanical assemblies (gyroscopes, etc.) cleaning and clothing dry cleaning (especially in Europe and Asia where CFC-113 is the primary dry cleaning fluid).

Table 6.9. CFC-113 Elimination Plans of Major Corporations.

Northern Telecom	1991
Mitsubishi	1991
Siemens	1991
Seiko-Epson	1993
American Telephone and Telegraph	1994
Fujitsu	1995
Sony	1995
Hitachi	1996
Sharp	1998
Nippon Electronics Corporation	2000
Toshiba	2000

6.8.2. Alternative CFC Cleaning Materials

In military and other high reliability cleaning applications, numerous substitution options for CFCs are currently available or are nearing final testing in preparation for commercialization. As Tables 6.10 and 6.11 indicate, all options have advantages and disadvantages that must be considered before selecting a CFC alternative cleaning material and its associated application equipment and processes. Methyl chloroform possibly will have limited applicability as a long-term option for replacing CFC-113 cleaning. However, methyl chloroform may have a viable role in the near future as an interim substitute for CFC-113 cleaning while chemical producers/users evaluate and implement long-term solutions before the turn of the century. Since methyl chloroform and CFC-113 along with their blends or azeotropic mixtures with alcohols are the primary solvent cleaning media readily allowed by current military cleaning specifications and standards Table 6.12, it would seem that major changes in the military's approach to

Table 6.10. Comparison of Cleaning Material Options versus CFC-113.

PROCESS	EQUIPMENT	SAFETY	PROVEN TECHNOLOGY	RECOVERABLE	DISPOSAL PROBLEM	ENERGY
CFC-113	Existing	Hi	Hi	Yes	Med	Lo/Med
Chlorinated	Existing	Med	Hi	Yes	Med	Med
Aqueous	Existing	Hi	Hi	No	Med	Hi
Semi-aqueous	New	Lo*	Lo	No	Med	Hi
Alcohol	New	Lo*	Med/Hi	Yes	Med	Med
HCFCs	New	Med/Hi#	Med/Hi	Yes	Med	Lo/Med

*Flammability issues.
#Toxicological testing not completed on all HCFC solvents.
Source: Allied Signal; table amended.

Table 6.11. Cleaning Technology Options.

OPTION	ADVANTAGES	DISADVANTAGES
CFC-113 conservation	Immediate use reductions	Maximum reduction 30–50%, equipment mods. likely
1,1,1-Trichloroethane	Readily available, existing equipment, existing technology	More aggressive than CFC-113, *will* be restricted by EPA, *will* be added to protocol
Aqueous (detergent)	Existing technology, existing equipment	Difficult to control when used as *only* cleaner, may required some component mounting/design changes
Semi-aqueous (terpenes)	Biodegradable, water rinse	New capital equipment not readily available, safety (flammability) issue, some hardware compatibility problems
Alcohol (IPA; IPA/water)	Readily available, existing technology	New capital equipment, safety (flammability) issues, regulatory VOC
HCFCs (14lb, etc.)	Slightly more aggressive than CFC-113	Still a low level ODP, safety (long term toxicity?) New capital equipment available '92–'93?

allowable cleaning materials and processes will require significant reassessment during the 1990s.

Alcohols are another class of cleaners that although allowed in limited quantities for small applications are experiencing a resurgence as possible CFC al-

Table 6.12. Primary Solvents Cleaners allowed by Military Electronics Specifications and Standards.

SOLVENT	SPECIFICATION
1,1,1-Trichloroethane	MIL-T-81533
Trichlorotrifluoroethane	MIL-C-81302
Trichlorotrifluoroethane	MIL-C-85447, Type II
Isopropyl alcohol	TT-I-735
Ethyl alcohol*	O-E-00760, Types III, IV, or V
Methyl alcohol*	O-M-232, Grade A
Butyl alcohol, secondary*	TT-B-848

*Allowed when added by supplier as part of a solvent blend. See WS-6536 and MIL-STD-2000.

Table 6.13. Comparison of HCFC Cleaners versus CFC-113.

DESIGNATION	FORMULA	B.P., °C	ATMOSPHERIC LIFETIME,[a] yrs	ODP[a]	GWP[a]	FLAMMABLE	TOXICOLOGY
CFC-113	CCl_2FCClF_2	47.6	90	0.83	1.4	No	Low
HCFC-123	$CHCl_2CF_3$	28.7	2	0.02	0.02	No	Incomplete
HCFC-141b	CCl_2FCH_3	32.0	10	0.09	0.1	Yes	Incomplete
HCFC-225ca	$CHCl_2CF_2CF_3$	51.1	—	<0.05	—	No	Incomplete
HCFC-225cb	$CHClFCF_2CClF_2$	56.1	—	<0.05	—	No	Incomplete
Genesolve 2010	*	29.6		0.07	0.08	No@	Incomplete
Genesolve 2020	**	31.6	—	0.06	0.07	No@	Incomplete
KCD-9434	#	30.0		0.07		No	Incomplete

*Allied-Signal Trademark for blend of: 86.1% (14lb), 10.0% (12), 3.6% (MeOH), 0.3% nitromethane.
**Allied-Signal Trademark for blend of: 80% (14lb), 20% (123), 0.1% nitromethane.
#Du Pont Trademark for blend of: 62.2% (14lb), 35.0% (123), 2.5% (MeOH), 0.3% stabilizer.
@However, has an upper explosion limit (UEL) and lower explosion limit (LEL) for ignition, as does methyl-chloroform.
[a]Period of time that the chlorine associated with a compound remains in the atmosphere.
[b]Ozone depleting potential [19].
[c]Global warming potential [19].

ternatives, especially in Europe. Commercial defluxing equipment in both batch and inline designs is available and has been successfully implemented in some countries to clean electronic assemblies. However, concerns over safety issues, (flammability) and the fact that alcohol vapors would fall under highly regulated VOC emissions may sevely limit the widespread use of this cleaning technology, especially in the United States, even though flammability safety factors can be effectively engineered into equipment designs.

Currently, the chemicals with the highest potential for becoming drop-in replacements for CFC-113 cleaning solvents are hydrochlorofluorocarbon (HCFC) solvents. HCFC-141b and its mixtures with other solvents (HCFC-123 and/or alcohols) have properties similar to CFC-113 although some modification of existing equipment may be required (Table 6.13). One of the most promising replacements, HCFC-225, is currently under development.[20] As Table 6.14 indicates, this three carbon HCFC has properties almost identical to CFC-113. The material however will probably not be available in commercial quantities until 1995–1997 at the earliest assuming short and long term toxicity testing results are acceptable. This HCFC has not only great promise for electronics cleaning but also represents the only viable replacement for CFC-113 dry cleaning operations.

Even though the HCFCs are more ozone friendly than CFCs, they are still ozone depleters and as such may have a relative short commercial life. Since the level of free chlorine is expected to rise (with a continued reduction in ozone) in the stratosphere for the next one to two decades as a result of the CFCs released

Table 6.14. Physical Properties of HCFC-225ca/225cb[22]

	HCFC-225ca, $CHCl_2CF_2CF_3$	HCFC-225cb, $CHClFCF_2CClF_2$	CFC-113, CCl_2FCClF_2
Molecular weight	202.94	202.94	187.38
Boiling point, °C	51.1	56.1	47.6
Freezing point, °C	−94.	−97.	−35.
Denisty g/cm³*	1.55	1.56	1.56
Viscosity, cP*	0.58	0.60	0.68
Surface tension dyne/cm*	15.8	16.7	17.3
Kauri-butanol value	34.	30.	31.
Flashpoint	None	None	None
Relative evaporation rate (ether = 100)	101.	84.	123.
Azeotropic Composition with EtOH, wt%	97.3/2.7	95.6/4.4	96.2/3.8

*At 25°C
#Table amended

over the last 10–20 years, the projected use of even low ozone depleting materials may be restricted and eventually legislated out of existence as is expected for CFCs. Recent predictions[21] have been made that HCFC solvents may have a potential commercial life of 30–40 years (i.e., 2020–2030) if the chlorine levels are to eventually be brought back to 1970 levels. This objective will probably not be reached until at least 2075 to 2100 even if CFCs and methyl chloroform are eliminated by the year 2000.

The use of naturally occurring materials such as terpenes and their derivatives are also possible substitutes for electronics CFC cleaners. This technology has been dubbed "semi-aqueous" cleaning since the application of terpene materials to remove contaminants is followed by water rinsing to remove the residual terpene. However these materials also have inherent flammability concerns similar to those for alcohol cleaning and like alcohols the safety concerns can be effectively removed by proper equipment design. The increased costs for equipment designed to negate these issues may limit the market penetration for these materials as well as for alcohols.

The most promising replacement for military and other high reliability electronics CFC solvent cleaning applications appears to be the use of aqueous cleaning. This technology, whether utilizing aqueous solutions of saponifiers to remove rosin based flux residues or rinse aids and water to remove nonrosin water soluble fluxes, has encountered wide spread use in the commercial sector and some limited use in military applications. With adequate component mounting considerations being incorporated into electronic assembly designs, aqueous cleaning can be utilized to produce high quality, high reliability electronics hardware. As such, this aqueous cleaning technology may provide an opportunity to move from CFC cleaning in a relative short time, certainly within 5–10 years. In addition, commercial equipment is presently available to utilize aqueous cleaning.

Although aqueous or semi-aqueous technologies appear at present to be viable ozone friendly cleaning alternatives, the issue of water pollution from aqueous cleaning effluent streams must be addressed. Even though these approaches use biodegradable constituents, the wholesale high volume use of these materials will almost certainly require some effluent pretreatment prior to release into local water treatment systems. The world cannot afford to trade atmospheric pollution for potential water pollution. (Refer to Chapter 4 for additional information on aqueous cleaning effluent concerns.)

6.9. CONCLUSIONS

Numerous alternative cleaning technologies are available that can replace most if not all current applications using CFC cleaning solvents. No one technology will probably replace all CFC applications, therefore evaluation testing will be required by most users. The major action remaining is to provide adequate data to the military and other high reliability customers supporting the efectiveness of these technologies. The task is not an easy one. Currently more than 49,000 military and Federal specifications and standards exist. In excess of 250 specifications and standards have been identified by the Department of Defense that directly specify the use of CFCs and Halons. Each document must be reviewed and the uses/applications specified will require close scrutiny before replacements are incorporated. With such information, end item requirements for both commercial and military hardware can be revised in a timely manner to allow the accelerated implementation of alternative, ozone friendly cleaning technologies in the electronics industry.

REFERENCES

1. DeNoon, R., and W. Hobson. 1972. "Printed Wiring Assemblies: Detection of Ionic Contaminants on." Report No. 3-72, Naval Avionics Facility.
2. Anon. 1988. "Surface Organic Contaminant Detection Test (In-House Method)." Test Method Number 2.3.38, Revision A. The Institute for Interconnecting and Packaging Electronic Circuits, February.
3. Anon. 1988. "Surface Organic Contaminant Identified Test Infrared Analytical Method". Test Method Number 2.3.39, Revision A. The Institute for Interconnecting and Packaging Electronic Circuits, February.
4. Wittmer, P., and B. Bonner. "Rational and Methodology for a Standard Residual Rosin Test by UV-Visible Spectrophotometric Methods." Magnavox Electronic Systems, Co. Fort Wayne, Indiana.

5. Anon. 1986. "Post Solder Aqueous Cleaning Handbook." ANSI/IPC-AC-62. The Institute for Interconnecting and Packaging Electronic Circuits, December.

6. Anon. 1987. "Post Solder Solvent Cleaning Handbook." ANSI/IPC-SC-60. The Institute for Interconnecting and Packaging Electronic Circuits, April.

7. Hayes, M. 1989. "Terpenes: A Halogen-Free CFC Alternative." Paper read at Proceedings of Electronics, CFCs and Stratospheric Ozone Conference, San Jose, California, February.

8. Hayes, M. 1988. "Chlorinated and CFC Solvent Replacement in the Electronics Industry: The Terpene Hydrocarbon Alternative." Petroferm, Inc., February.

9. Johnson, K., and D. Sanger. 1983. "A Study of Solvent and Aqueous Cleaning of Fluxes," NWC-TP-6427. Naval Weapons Center, China Lake, California.

10. DeNoon, R., and W. Hobson. 1978. "Review of Data Generated with Instruments Used to Detect and Measure Ionic Contaminants in Printed Wiring Assemblies." Materials Research Report Number 3-78, Naval Avionics Center.

11. Gorondy, Emery I. 1984. "Surface Insulation Resistance—Part I: The Development of an Automated SIR Measurement Technique," IPC-TP-518.

12. Gorondy, Emery I. 1985. "Surface Insulation Resistance—Part II: Exploring the Correlation Between Standard Industry and Military "SIR" Test Patterns—A Status Report." IPC-TP-543.

13. Gorondy, Emery I. 1988. "Surface/Moisture Insulation Resistance (SIR/MIR)—Part III: Analysis of the Effect of Test Parameters and Environmental Conditions on Test Results." Presented at IPC Conference, Anaheim, California, October.

14. Chan, A. S. L., and T. A. Shankoff. 1988. "A Correlation between Surface Insulation Resistance and Solvent Extract Conductivity Cleanliness Tests." *Circuit World*, **14** (November), 23–26.

15. Archer, W., T. Cabelka, and J. Nalazek. "Quantitative Determination of Rosin Residues on Cleaned Electronic Assemblies," NWC-TP-6789. Available from Naval Weapons Center, Ridgecrest, California.

16. Wittmer and Bonner, op. cit.

17. Anon. 1987. Montreal Protocol on Substances that deplete the Ozone Layer (Final Act) United Nations Environmental Program.

18. U. S. National Aeronautics and Space Administration. 1988. "Present State of Knowledge of the Upper Atmosphere 1988: An Assessment Report." Washington, D. C., August.

19. Anon. 1989. "Scientific Assessment of Stratospheric Ozone: 1989." Appendix AFEAS Report, United Nations Environmental Program/World Meterological Organization, September.

20. Yamabe, M. 1989. "HCFC-225ca and 225cb as Alternatives to CFC-113." Presented at The International Conference on CFC and Halon Alternatives, Washington, D.C., October.

21. Anon. 1989. International Conference on CFC and Halon Alternatives Washington, D.C., October 10–11.

22. Ibid.

BIBLIOGRAPHY

This listing is arranged alphabetically by author's name. It was compiled from those articles in the files of the editor which related to the subject of the book, and supplements the references cited in the various chapters. Failure to include any given article or paper does not imply anything except that the editor either did not read it, was not aware of it, or felt it duplicated information already included.

Agnew, J. "Choose Cleaning Solvents Carefully." **Elec. Dsg.**, **5**(1), 54–57 (March 1974).

Anderson, S. "Stratospheric Ozone: Federal Regulations and the Use of Chlorofluorocarbons in the Electronics Industry." NWC TP 6896.

Anderson, S. "The Future of CFCs in Electronic Cleaning—A Regulatory Update." *Proc. NEPCON WEST,* 1593–1600 (Feb. 1990).

Anon. MIL-F-14256D. "Flux, Soldering, Liquid (Rosin Base)." (17 April 1972).

Anon. "Printed-Wiring Assemblies: Detection of Ionic Contaminants on." Naval Avionics Center Mat. Res. Report No. 3-72 (25 May 1972).

Anon. MIL-P-28809. "Printed-Wiring Assemblies." (21 March 1975).

Anon. MIL-F-14256D, Amend 1. "Flux, Soldering, Liquid (Rosin Base)." (12 June 1975).

Anon. "Review of Data Generated with Instruments used to Detect and Measure Ionic Contaminants on Printed-Wiring Assemblies." Naval Avionics Center Mat. Res Report No. 3-78 (29 Aug. 1978).

Anon. Deleterious Effect of MIL-F-14256, Type RA Fluxes on Printed Wiring Boards." Naval Avionics Center Tech. Report No. 2253, (31 Jan. 1979).

Anon. "Factors Affecting Insulation Resistance Performance of Printed Boards." IPC Raw Mat. Com. Report No. 468 (March 1979).

Anon. "Deleterious Effect of Fusing Fluids on Printed Wiring Boards." Naval Avionics Center Tech. Report No. 2259 (9 May 1979).

Anon. "General Requirements for Soldering Electronic Interconnections." IPC Std. Spec. IPC-S-815A (Aug. 1980).

Anon. *Printed Wiring and Backplane Industry Forecast 1982*. Menlo Park, CA: Gnostic Concepts, 1982.

Anon. "An International Soldering Flux Specification for the Electronics Industry." *The Western Electric Engineer* (First Issue), 15–21 (1983).

Anon. *Surface Mount Technology*, Technical Monograph Series 6984-002. Silver Spring, MD: ISHM, 1984.

Anon. "Surface Mount Land Patterns (Configurations and Design Rules)." ANSI/IPC-SM-782 (1987).

Anon. "Guidelines for Cleaning of Printed Boards and Assemblies." IPC-CH-65 (1990).

Archer. W. L., "Examination of Solvent Cleaning For Printed Wiring Board Applications." *Electri-Onics,* 11–14 (May 1987).

Archer, W. L., T. D. Cabelka, and J. J. Nalazek. "The Problem of Residual Flux." *Ass. Eng.* 20–23 (Dec. 1987).

Attalla, G. "In-Line Semi-Aqueous Cleaning." *Proc. NEPCON WEST,* 562–565 (Feb. 1990).

Ball, D. F. "Aqueous Flux Removal After IR Reflow—Equipment and Fluxes." Chemcut Corp. n.d.

Barber, C. L. *Solder—Its Fundamentals and Usage,* 3rd ed. Chicago: Kester Solder Co., 1965.

Basu, R. S., and J. K. Bonner. "Alternatives to CFCs: New Solvents for the Electronics Industry." *Surface Mount Technol.* 34–37 (Dec. 1989).

Bester, M. H., and C. B. Titus, Jr. "Cost Effectiveness of Some Organic Solvent and Aqueous Cleaning Procedures." Hughes Aircraft Co. Test Report No. 80-2-26 (26 Feb. 1980).

Basu, R. S., and J. K. Bonner. "A Quasi-Empirical Method for Predicting the Dissolution of Flux Residues on Printed Wiring Assemblies." IPC Tech. Paper No. 649.

Bonner, J. K. "Establishment of Production Cleanliness Criteria and Processes for Printed Wiring Boards and Assemblies: Phase I." Final Tech. Rept., Contract No. DAAK40-78-C-0114 (April 1979).

Bonner, J. K. "Review of the MIRADCOM Cleanliness Criteria Questionnaire for Printed Wiring." IPC Tech. Paper No. 301 (Sept. 1979).

Bonner, J. K. "Ultrasonic Cleaning for Printed Wiring Assemblies." Martin Marietta Orlando Aerospace IRAD Rept., (Dec. 1981). Unpublished.

Bonner, J. K. "Establishment of Production Cleanliness Criteria and Processes for Printed Wiring Boards and Assemblies: Phase II." Final Tech. Rept., Contract No. DAAK40-78-C-0114 (Dec. 1981).

Bonner, J. K. "A Critical Comparison of Cleaning Different Fluxes with Different Cleaning Techniques." *Proc. Nepcon West '85* (Feb. 1985).

Bonner, J. K. "A New Solvent for Post-Solder Cleaning of Printed Wiring Assemblies." *Proc. Nepcon West '86* (Feb. 1986).

Bonner, J. K. "Automation and Cleaning: A New Opportunity for SMT." *Proc. SMART IV* (Jan. 1988).

Bonner, J. K. and V. A. De Giorgio. "A Systems Approach for Detecting Contaminant Species on Printed Wiring Boards." IEPS Tech. Paper (Nov. 1981).

Bonner, J. K. and H. F. Osterman. "A New Process for Cleaning Surface Mount Assemblies." *Proc. Nepcon West '87* (Feb. 1987).

Brands, Edwin R. "The Effect of Activated Flux and Cleaning Methods on Performance of PCB's in Humidity." *Fifth Cont. Control Sem. Paper* (16–18 Oct. 1979).

Breunsbach, R., "Semi-Aqueous Batch Cleaning of SMD's." *Proc. NEPCON WEST,* 553–561 (Feb. 1990).

Brous, J. "Effects of Water Soluble Flux Materials on Circuit Board Insulation Resistance." Alpha Metals Tech. Paper, n.d.

Bud, Paul J. "Solderability Assurance and Corrosion Protection of PWB's and Components, Part 3." *Ins./Circ.,* 31–35 (Oct. 1980).

Caci, S. F., and W. P. Mikelonis. "Soldering with Water Soluble Flux and Aqueous Chemistry Flux Removal." IPC Tech. Paper No. 246 (Sept. 1978).

Capillo, C. A. "Surface Mount Assemblies Create New Cleaning Challenges." Elec. Pack. & Prod., 76–81 (Aug. 1984).

Casperson, P. G. "The Nature of Residues Presented to the Cleaning Machine." Elec. Pack. & Prod., 73–79 (Jan. 1980).

Chung, B. C., and J. C. Eagle, "Detection of Organic Components in Non-Rosin Fluxes." Western Electric Co. Report, n.d.

Connelley, J. A. "Solving the Ozone Problem Before it's Too Late." *Electronic Business,* 72–73 (Feb. 20, 1989).

Cooley, Dwight. "Evaluation of Cleaning Processes After Solder Fuzing." IPC Tech. Paper No. 102 (April 1976).

Coombs, C. F., Jr. (Ed.). *Printed Circuits Handbook,* 1st ed. New York: McGraw-Hill, 1967.

Coombs, C. F. *Printed Circuits Handbook,* 2nd ed. New York: McGraw-Hill, 1979.

Dallessandro, S. M., and T. D. Cabelka. "Identifying Causes and Effects of Ionic Contamination in PWA Assemblies—Part 2." *Electri-Onics,* 28–30 (Oct. 1985).

Denison, J. W., Jr. "Cleaning of Printed Circuit Boards to Remove Ionic Soils." Corrosion/74, Paper No. 27 (4–8 March 1974).

Dishart, K. T., and M. C. Wolff. "Advantages and Process Options of Hydrogen Based Formulations in Semi-Aqueous Cleaning." *Proc. NEPCON WEST,* 513–520 (Feb. 1990).

Dukes, J. M. C. *Printed Circuits: Their Design and Application.* London: MacDonald, 1961.

Duyck, T. O. "Testing Large Printed Circuit Boards for Cleanliness." *Ins. Circ.,* 38–41 (Oct. 1978).

Engelland, G. J. and W. G. Kenyon. "Insulation Resistance Degradation by Non-Ionic Surface Contamination: Its Cause and Cure." IPC Tech. Paper No. 135 (Sept. 1976).

Engelland, G. J., and W. G. Kenyon. "Non-Ionic Surface Contamination Degrades Insulation Resistance." *Circ. Mfg.,* 34–40 (June 1977).

Figiel, F. J. "Cleaning PCB Assemblies: Nonazeotropic Solvent Mixture Offers an Alternative." *Circ. Mfg.,* 80–85 (Oct. 1979).

Geckle, R. A. "Cleaning Procedures and Solvents for Semiconductors and PC Boards." *Elec. Pack. & Prod.,* 127–138 (July 1975).

Gillman, A. "Methods of Solder Flux Removal." *Elec. Pack. & Prod.,* 137–141 (Jan. 1979).

Giroux, J. "Defluxing with Chlorinated Solvent: Better than Aqueous Systems?" *Circ. Mfg.,* 75–77 (Feb. 1981).

Harper, C. A. (Ed.). *Handbook of Electronic Packaging.* New York: McGraw-Hill, 1969.

Harrington, J., Jr. *Understanding the Manufacturing Process.* New York: Marcel Dekker, 1984.

Hasson, J. C. and M. V. Kulkarni. "Open-Tube-Column Gas Chromatography of Rosin Fluxes." *Anal. Chem.,* **44**(9), 1586–1589 (Aug. 1972).

Hayes, M. E. "Cleaning SMT Assemblies Without Halogenated Solvents." *Surface Mount Technology,* 37–40 (Dec. 1988).

Hayes, M. "Advances and Acceptance of Electronic Assembly Cleaning with Terpenes." *Proc. NEPCON WEST,* 499–512 (Feb. 1990).

Heuring, H. F. "Cleaning Electronic Components and Subassemblies." *Elec. Pack. & Prod.,* 10–14 (June 1967).

Howell, D. "Make P-C Boards Come Clean." *Elec. Prod. Mag.,* 27–34 (June 1976).

Hudy, J. A. "Resin Acids—Gas Chromatography of Their Methyl Esters." *Anal. Chem.,* **31**(11) 1754–1756 (Nov. 1959).

Hume, W. A. "Proper Cleaning of Electronic Assemblies." *Ass. Eng.,* 36–42 (March 1968).

Hutchins, C. L. *Surface Mounting Technology—How to Get Started.* Warrington, PA: INFO-MATION, 1986.

Jennings, C. W. "Measurement of Electrical Properties Determines Actual Board Capabilities." *Ins./Circ.,* 29–40 (Feb. 1978).

Kear, F. W. *Printed Circuit Assembly Manufacturing.* New York: Marcel Dekker, 1987.

Keeler, R. "Aqueous Cleaning Challenges Solvent in Flux Removal." *Elec. Pack. & Prod.,* 50–55 (Aug. 1984).

Keller, J. "Improving PWB Reliability While Cutting Soldering and Cleaning Costs—Part 3." *Ass. Eng.,* 18–21 (July 1981).

Kenyon, W. G. "Fundamentals of Ionic Contamination Measurement Using Solvent Extract Methods." IPC Tech. Paper No. 177 (Sept. 1977).

Kenyon, W. G. "Improved RA Flux Cleaning with *New Fluorocarbon/Alcohol Combination.*" *Nepcon West '77 Tech. Paper.*

Kenyon, W. G. "Part 1—Water Cleaning Assemblies: Wave of the Future or Washout?" *Ins./Circ.* (Feb. 1978).

Kenyon, W. G. "How to Use the Solvent Extract Method to Measure Ionic Contamination of Printed Wiring Assemblies." *Ins./Circ.*, 47–50 (March 1981).

Kenyon, W. G. and J. J. Daly, Jr. "A Comparison of Solvent and Aqueous Processes for Cleaning Printed Wiring Assemblies." *Internepcon '78,* Tech. Rept. RP-8 (Sept. 1979).

Kenyon, W. G., and D. S. Lermond. "Post-Solder Cleaning of SMAs." *Printed Wiring Assembly,* 20 (Aug. 1987).

Kolman, Frank L. "Systemizing Identification of Equipment Requirements/Selection for Solder/Deflux System." *Electri-Onics,* 49–52 (July 1983).

Lacy, M. E. "Isolation and Molecular Identification of Ultramicro Contaminants by Fourier Transform Infrared Spectroscopy." *Fifth Cont. Control Sem. Papers,* (16–18 Oct. 1979).

Langan, J. P. "How to Select Fluxes for Maximum Effectiveness." *Ins./Circ.,* 79–80 (Dec. 1978).

Leonida, G. *Handbook of Printed Circuit Design, Manufacture, Components and Assembly.* Ayr, Scotland: Electrochemical Publications Ltd., 1981.

Licari, J. J. *Plastic Coatings for Electronics.* New York: McGraw-Hill, 1970.

Licari, J. J. "Materials, Methods and Equipment for Cleaning Electronic Assemblies." *Ins./Circ.,* 35–39 (May 1970).

Lyman, J. (Ed.). *Microelectronic Interconnection and Packaging.* New York: McGraw-Hill, 1980.

MacKay, C. A. "What You Don't Know about Solder Creams." *Circ. Mfg.,* 43–52 (May 1987).

Manko, H. H. "New Packaging Techniques force a Reexamination of Cleaning Methods." *Elec. Pack. & Prod.,* 68–73 (Aug. 1984).

Manko, H. H. "Advantages and Pitfalls of 'No Clean' Flux Systems." *Proc. NEPCON WEST,* 1093–1104 (Feb. 1990).

Marczi, M., N. Bandyopadhay, and S. Adams. "No-Clean, No-Residue Soldering Process." *Circ. Mfg.,* 42–46 (Feb. 1990).

Markstein, H. W. "Solder Flux Developments Expand Choices." *Elec. Pack. & Prod.,* 39–42 (April 1983).

Matisoff, B. S. *Handbook of Electronics Manufacturing Engineering.* New York: Van Nostrand Reinhold, 1978.

Merchant, A. and M. C. Wolff. "Evaluating New, 'Safe' Cleaners and Defluxers." *Circ. Mfg.,* 46–51 (Nov. 1989).

Mindel, M. J. "Flux Corrosion Test Reflects Accurate Field Operating Conditions." *Elec. Pack. & Prod.,* 76–77 (Dec. 1989).

Morris, J. R., and N. Bandyopadhyay. "No-Clean Solder Paste Reflow Processes." *Printed Circuit Assembly,* 26–31 (Feb. 1990).

Murphy, D. "Cleaning Printed Circuit Boards—Experts Review the Problem." *Circ. Mfg.,* 52–66 (July 1974).

Phillips, H. E. "Ionic Residue Removal: Which Solvent Is Best?" *Elec. Pack. & Prod.,* 177–180 (Sept. 1973).

Polhamus, R. L. "Cleaning Beneath Surface Mount Devices Using MicroCoustic Technology." *Surface Mount Technology,* 29–31 (Dec. 1989).

Poteet, D. "Insulation Resistance, PWB Materials and Processes." IPC Tech. Paper No. 85 (April 1976).

Pound, R. "Solvents Flush Solder Flux from Tight Spaces." *Elec. Pack. & Prod.,* 58–63 (Aug. 1984).

Quina, J., and S. Wagner. "Contamination and Cleaning in Electronics Manufacture and Assembly." *Fifth Cont. Control Sem. Papers* (16–18 Oct. 1979).

Raffalovich, A. J. "Corrosion Effects of Solder Flux on Printed-Circuit Boards." *IEEE Trans./Parts, Hybrids, and Pack.*, **PHP 7**(4), 155–162 (Dec. 1971).

Randolph. S. "How Clean is Centrifugal Clean?" *Circ. Mfg.*, 90–94 (Feb. 1990).

Rigling, Walter S. and Lish, Earl F. "An Analysis of Polar and Non Polar PC Assembly Contamination Removal." *Internepcon '77 Papers*, 4–7 (18–20 Oct. 1977).

Schneider, A. F., "How to Evaluate and Select Solvent Cleaners for Rosin Flux Removal." SME Tech Paper No. EE77-863 (1977).

Schneider, A. F. "Water-Soluble Organic Fluxes for the Electronics Industry." *Ins./Circ.*, 31–36 (Feb. 1979).

Schuessler, P. "Water-Soluble Flux Technology." *Inst./Circ.*, 88–91 (Dec. 1981).

Shoemaker, S., L. Fisher, and L. Crane. "Ionic Contaminant Cleaning of Dry Film Solder Masked High Reliability Boards." *Printed Wiring Assembly*, 17–20 (Oct. 1987).

Sinnadurai, F. N. (Ed.). *Handbook of Microelectronics Packaging and Interconnection Technologies*. Ayr, Scotland: Electrochemical Publications Ltd., 1985.

Slezak, E. "Considerations for Water Soluble Solderpaste Use." *Proc. NEPCON WEST*, 1609–1617 (Feb. 1990).

Slinn, D. S. L., and B. H. Baxter. "Alcohol Cleaning Under A Non-Flammable Perfluorocarbon Vapor Blanket." *Proc. NEPCON WEST*, 1810–1819 (Feb. 1990).

Small, E. "No-Clean Fluxes: New Technology Needs New Test," *Circ. Mfg.*, 52–56 (Dec. 1989).

Soble, R. M. "Solvent Cleaning of Printed Wiring Assemblies." *Ins./Circ.*, 25–29 (Oct. 1979).

Spitz, S. L. "Cleaning Printed Circuit Boards for Higher Quality." *Elec. Pack. & Prod.* 100–106 (Sept. 1985).

Tadros, M., T. Shah, and C. Matzdorf. "A Safe and Effective Cleaner for Printed Wiring Boards." *Proc. NEPCON WEST*, 1622–1625 (Feb. 1990).

Tautscher, C. J. "Causes and Prevention of Blisters in Conformal Printed Circuit Coatings." *Ins./Circ.*, 32–33 (June 1972).

Tautscher, C. J. "PCB Contamination Introduced by the Soldering Process." *Ins./Circ.*, 9–18 (July 1978).

Tautscher, C. J. "New Generation of Board and Assembly Cleanliness Tests." *Ins./Circ.*, 15–19 (Dec. 1979).

Tautscher, C. J. "Influencing Factors for Ionic Surface Cleanliness of PWB Assemblies." E-Systems/ECI Div. Paper (ca. 1980).

Tautscher, C. J. "Cleanliness Testing, Present Approaches, Trends, and Needs for Improvement." Presented at the IPC, April (1981).

Taylor, P. R., P. L. Altavilla, and G. J. Simons. "Solvent Extract Resistivity Testing—Is It Meaningful?" IPC Tech. Paper No. 41 (Sept. 1974).

Treiber, J. "Cleaning SMAs: Water vs. Solvent." *Surface Mount Technology*, 45.

Trombka, J. A. "Solvent Solutions." *Circ. Mfg.*, 45–48 (Nov. 1985).

Turbini, L. J. "CFC Alternatives." *Printed Circuit Assembly*, 16–18 (Feb. 1990).

Turbini, L. J., J. G. Eagle, and T. J. Stark. "A Comparison of Removal of Activated Rosin Flux by Selected Solvents." Western Electric Co. Paper, n.d.

U.S. EPA. "Analysis of the Environmental Implications of the Future Growth in Demand for Partially-Halogenated Chlorinated Compounds." Peer Review Draft, Office of Air and Radiation, U.S. Environmental Protection Agency, Washington, DC 20460 (July 24, 1989).

Wagner, S. "Copper (II) Abietate, Basic Copper (II) Abietate." *Honeywell Test Report* No. 15165 (14 May 1975).

Wargotz, W. B. "Quantification of Contamination Effects Upon Electrical Behavior of Printed Wiring." IPC Tech. Paper No. 182 (Sept. 1977).

Wargotz, W. B. "Cleanliness and Reliability: Quantifying Contaminant Effects." *Circ. Mfg.*, 40–49 (1978).

Willard, H. H., L. L. Merritt, Jr., J. A. Dean, and F. A. Settle, Jr. *Instrumental Methods of Analysis*, 6th ed. Belmont, CA: Wadsworth Publishing, 1981.

Wittmer, P. "Optimizing Materials and Process Variables to Aid Cleaning." *Proc. Nepcon West '87* (Feb. 1987).

Wun, K.-L. W., and J. Baker, "Non-CFC Cleaning and Cleanliness Testing for Printed Circuit Assemblies, Part 1." *Surface Mount Technology*, 39–44 (Dec. 1989).

Wun, K.-L. W., and J. Baker, Jim, "Non-Cleaning and Cleanliness Testing for Printed Circuit Assemblies, Part 2." *Surface Mount Technology*, 49–54 (Jan. 1990).

Young, J. A. "Automated Cleaning of SMD Assemblies.' *Ass. Engl.* (June 1984).

Zado, F. M. "Insulation Resistance Degradation Caused by Nonionic Water Soluble Flux Residues." *Nepcon West '79 Tech. Papers*, 346–353 (March 1979).

Zado, F. M. "Interrelationships in the System: Ionic Residues/Insulation Resistance/PWB Assembly Reliability." Western Electric Co. Report, n.d.

Index